Waves and Optics Simulations
The Consortium for
Upper-Level Physics Software

Wolfgang Christian, Andrew Antonelli, and Susan Fischer
Department of Physics, Davidson College,
Davidson, North Carolina

Robin A. Giles
Department of Physics and Astronomy, Brandon University,
Brandon, Manitoba, Canada

Brian W. James
Department of Physics, University of Salford,
Salford, United Kingdom

Ronald Stoner
Department of Physics, Bowling Green University,
Bowling Green, Ohio

Series Editors

William MacDonald

Maria Dworzecka

Robert Ehrlich

JOHN WILEY & SONS, INC.

NEW YORK · CHICHESTER · BRISBANE · TORONTO · SINGAPORE

ACQUISITIONS EDITOR Cliff Mills
MARKETING MANAGER Catherine Faduska
PRODUCTION EDITOR Sandra Russell
MANUFACTURING MANAGER Mark Cirillo

This book was set in 10/12 Times Roman by Beacon Graphics and
printed and bound by Hamilton Printing Co. The cover was printed by Phoenix Color.

Library of Congress Cataloging in Publication Data:
Christian, Wolfgang, Antonelli, Andrew, Fischer, Susan, Giles, Robin, James, Brian, Stoner, Ronald
    Waves and Optics Simulations: the consortium for upper level
  physics software / MacDonald, William, Dworzecka, Maria, Ehrlich, Robert
    Includes bibliographical references (p.).
    ISBN 0-471-54887-1 (pbk./disk)

Printed in the United States of America

10 9 8 7 6 5 4 3 2

Contents

List of Figures

List of Tables

1

Introduction

"It is nice to know that the computer understands the problem. But I would like to understand it too."

—Eugene P. Wigner, quoted in *Physics Today,* July 1993

1.1 Using the Book and Software

The simulations in this book aim to exploit the capabilities of personal computers and provide instructors and students with valuable new opportunities to teach and learn physics, and help develop that all-important, if somewhat elusive, physical intuition. This book and the accompanying diskettes are intended to be used as supplementary materials for a junior- or senior-level course. Although you may find that you can run the programs without reading the text, the book is helpful for understanding the underlying physics, and provides numerous suggestions on ways to use the programs. *If you want a quick guided tour through the programs, consult the "Walk Throughs" in Appendix A.* The individual chapters and computer programs cover mainstream topics found in most textbooks. However, because the book is intended to be a supplementary text, no attempt has been made to cover all the topics one might encounter in a primary text.

Because of the book's organization, students or instructors may wish to deal with different chapters as they come up in the course, rather than reading the chapters in the order presented. One price of making the chapters semi-independent of one another is that they may not be entirely consistent in notation or tightly cross-referenced. Use of the book may vary according to the taste of the student or instructor. Students may use this material as the basis of a self-study course. Some instructors may make homework assignments from the large number of exercises in each chapter or to use them as the basis of student projects. Other instructors may use the computer programs primarily for in-class demonstrations. In this latter case, you may find that the programs are suitable for a range of courses from the introductory to the graduate level.

Use of the book and software may also vary with the degree of computer programming performed by users. For those without programming experience, all the computer simulations have been supplied in executable form, permitting them to be used as is. On the other hand, Pascal source code for the programs has also been provided, and a number of exercises suggest specific ways the programs can be modified. Possible modifications range from altering a single procedure especially set up for this purpose by the author, to larger modifications following given examples, to extensive additions for ambitious projects. However, the intent of the authors is that the simulations will help the student to develop intuition and a deeper understanding of the physics, rather than to develop computational skills.

We use the term "simulations" to refer to the computer programs described in the book. This term is meant to imply that programs include complex, often realistic, calculations of models of various physical systems, and the output is usually presented in the form of graphical (often animated) displays. Many of the simulations can produce numerical output—sometimes in the form of output files that could be analyzed by other programs. The user generally may vary many parameters of the system, and interact with it in other ways, so as to study its behavior in real time. The use of the term simulation should not convey the idea that the programs are bypassing the necessary physics calculations, and simply producing images that look more or less like the real thing.

The programs accompanying this book can be used in a way that complements, rather than displaces, the analytical work in the course. It is our belief that, in general, computational and analytical approaches to physics can be mutually reinforcing. It may require considerable analytical work, for example, to modify the programs, or really to understand the results of a simulation. In fact, one important use of the simulations is to suggest conjectures that may then be verified, modified, or proven false analytically. A complete list of programs is given in Section 1.7.

1.2   Required Hardware and Installation of Programs

The programs described in this book have been written in the Pascal language for MS-DOS platforms. The language is Borland/Turbo Pascal, and the minimum hardware configuration is an IBM-compatible 386-level machine preferably with math coprocessor, mouse, and VGA color monitor. In order to accommodate a wide range of machine speeds, most programs that use animation include the capability to slow down or speed up the program. To install the programs, place disk number 1 in a floppy drive. Change to that drive, and type Install. You need only type in the file name to execute the program. Alternatively, you could type the name of the driver program (the same name as the directory in which the programs reside), and select programs from a menu. A number of programs write to temporary files, so you should check to see if your autoexec.bat file has a line that sets a temporary directory, such as SET TEMP = C:\TEMP. (If you have installed WINDOWS on your PC, you will find that such a command has already been written into your autoexec.bat file.) If no such line is there, you should add one.

Compilation of Programs

If you need to compile the programs, it would be preferable to do so using the Borland 7.0 (or later) compiler. If you use an earlier Turbo compiler you may run out of memory when compiling. If that happens, try compiling after turning off memory resident programs. If your machine has one, be sure to compile with the math-coprocessor turned on (no emulation). Finally, if you recompile programs using any compiler other than Borland 7.0, you will get the message: "EGA/VGA Invalid Driver File" when you try to execute them, because the driver file supplied was produced using this version of the compiler. In this case, search for the file BGILINK.pas included as part of the compiler to find information on how to create the EGAVGA.obj driver file. *If any other instructions are needed for installation, compilation, or running of the programs, they will be given in a README file on the diskettes.*

1.3   User Interface

To start a program, simply type the name of the individual or driver program, and an opening screen will appear. All the programs in this book have a common user interface. Both keyboard and mouse interactions with the computer are possible. Here are some conventions common to all the programs.

Menus: If using the *keyboard*, press **F10** to highlight one of menu boxes, then use the **arrow** keys, **Home**, and **End** to move around. When you press **Return** a submenu will pull down from the currently highlighted menu option. Use the same keys to move around in the submenu, and press **Return** to choose the highlighted submenu entry. Press **Esc** if you want to leave the menu without making any choices.

If using the *mouse* to access the top menu, click on the menu bar to pull down a submenu, and then on the option you want to choose. Click anywhere outside the menus if you want to leave them without making any choice. Throughout this book, the process of choosing submenu entry **Sub** under main menu entry **Main** is referred to by the phrase "choose **Main | Sub**." The detailed structure of the menu will vary from program to program, but all will contain **File** as the first (left-most) entry, and under **File** you will find **About CUPS**, **About Program**, **Configuration**, and **Exit Program**. The first two items when activated by mouse or arrows keys will produce information screens. Selecting **Exit Program** will cause the program to terminate, and choosing **Configuration** will present you with a list of choices (described later), concerning the mode of running the program. In addition to these four items under the **File** menu, some programs may have additional items, such as **Open**, used to open a file for input, and **Save**, used to save an output file. If **Open** is present and is chosen, you will be presented with a scrollable list of files in the current directory from which to choose.

Hot Keys: Hot keys, usually listed on a bar at the bottom of the screen, can be activated by pressing the indicated key or by clicking on the hot key bar with the mouse. The hot key **F1** is reserved for help, the hot key **F10** activates the menu bar. Other hot keys may be available.

Sliders (scroll-bars): If using the *keyboard*, press **arrow** keys for slow scrolling of the slider, **PgUp/PgDn** for fast scrolling, and **End/Home** for moving from one end to another. If you have more then one slider on the screen then only the slider with marked "thumb" (sliding part) will respond to the above keys. You can toggle the mark between your sliders by pressing the **Tab** key.

If using the *mouse* to adjust a slider, click on the thumb of the slider, drag it to desired value, and release. Click on the arrow on either end of the slider for slow scrolling, or in the area on either side of thumb for fast scrolling in this direction. Also, you can click on the box where the value of the slider is displayed, and simply type in the desired number.

Input Screens: All input screens have a set of "default" values entered for all parameters, so that you can, if you wish, run the program by using these original values. Input screens may include circular radio buttons and square check boxes, both of which can take on Boolean, i.e., "on" or "off," values. Normally, check boxes are used when only one can be chosen, and radio buttons when any number can be chosen.

If using the *keyboard*, press **Return** to accept the screen, or **Esc** to cancel it and lose the changes you may have made. To make changes on the input screen by keyboard, use **arrow** keys, **PgUp**, **PgDn**, **End**, **Home**, **Tab**, and **Shift-Tab** to choose the field you want to change, and use the backspace or delete keys to delete numbers. For Boolean fields, i.e., those that may assume one of two values, use any key except those listed above to change its value to the opposite value.

If you use the *mouse*, click [OK] to accept the screen or [Cancel] to cancel the screen and lose the changes. Use the mouse to choose the field you want to change. Clicking on the Boolean field automatically changes its value to the opposite value.

Parser: Many programs allow the user to enter expressions of one or more variables that are evaluated by the program. The function parser can recognize the following functions: absolute value (abs), exponential (exp), integer or fractional part of a real number (int or frac), real or imaginary part of a complex number (re or im), square or square root of a number (sqr or sqrt), logarithms—base 10 or e (log or ln)—unit step function (h), and the sign of a real number (sgn). It can also recognize the following trigonometric functions: sin, cos, tan, cot, sec, csc, and the inverse functions: arcsin, arccos, arctan, as well as the hyperbolic functions denoted by adding an "h" at the end of all the preceding functions. In addition, the parser can recognize the constants $pi, e, i(\sqrt{-1})$, and rand (a random number between 0 and 1). The operations **+**, **−**, *****, **/**, **^** (exponentiation), and **!**(factorial) can all be used, and the variables r and c are interpreted as $r = \sqrt{x^2 + y^2}$ and $c = x + iy$. Expressions involving these functions, variables, and constants can be nested to an arbitrary level using parentheses and brackets. For example, suppose you entered the following expression: **h(abs(sin(10*pi* x))−0.5)**. The parser would interpret this function as $h(|sin(10\pi x)| - 0.5)$. If the program evaluates this function for a range of x-values, the result, in this case, would be a series of square pulses of width 1/15, and center-to-center separation 1/10.

Help: Most programs have context-sensitive help available by pressing the **F1** hot
key (or clicking the mouse in the **F1** hot key bar). In some programs help
is available by choosing appropriate items on the menu, and in still other
programs tutorials on various aspects of the program are available.

1.4   The CUPS Project and CUPS Utilities

The authors of this book have developed their programs and text as part of the
Consortium for Upper-Level Physics Software (CUPS). Under the direction of the
three editors of this book, CUPS is developing computer simulations and associated texts for nine junior- or senior-level courses, which comprise most of the undergraduate physics major curriculum during those two years. A list of the nine
CUPS courses, and the authors associated with each course, follows this section.
This international group of 27 physicists includes individuals with extensive back-
grounds in research, teaching, and development of instructional software.

The fact that each chapter of the book has been written by a different author
means that the chapters will reflect that individual's style and philosophy. Every
attempt has been made by the editors to enhance the similarity of chapters, and
to provide a similar user interface in each of the associated computer simulations. Consequently, you will find that the programs described in this and other
CUPS books have a common look and feel. This degree of similarity was made
possible by producing the software in a large group that shared a common philosophy and commitment to excellence.

Another crucial factor in developing a degree of similarity between all CUPS
programs is the use of a common set of utilities. These CUPS utilities were written
by Jaroslaw Tuszynski and William MacDonald, the former having responsibility
for the graphics units, and the latter for the numerical procedures and functions.
The numerical algorithms are of high quality and precision, as required for reli-
able results. CUPS utilities were originally based on the M.U.P.P.E.T. utilities of
Jack Wilson and E.F. Redish, which provided a framework for a much expanded
and enhanced mathematical and graphics library. The CUPS utilities (whose
source code is included with the simulations with this book), include additional
object-oriented programs for a complete graphical user interface, including pull-
down menus, sliders, buttons, hot-keys, and mouse clicking and dragging. They
also include routines for creating contour, two-dimensional (2-D) and 3-D plots,
and a function parser. The CUPS utilities have been provided in source code form
to enable users to run the simulations under future generations of Borland/Turbo
Pascal. If you do run under future generations of Turbo or Borland Pascal on the
PC, the utilities and programs will need to be recompiled. You will also need
to create a new egavga.obj file which gets combined with the programs when an
executable version is created—thereby avoiding the need to have separate
(egavga.bgi) driver files. These CUPS utilities are also available to users who
wish to use them for their own projects.

One element not included in the utilities is a procedure for creating hard copy
based on screen images. When hard copy is desired, those PC users with the appropriate graphics driver (graphics.com), may be able to produce high-quality
screen images by depressing the **PrintScreen** key. If you do not have the graphics software installed to get screen dumps, select **Configuration l Print Screen**,

and follow the directions. Moreover, public domain software also exists for capturing screen images, and for producing PostScript files, but the user should be aware that such files are often quite large, sometimes over 1 MB, and they require a PostScript printer driver to produce.

One feature of the CUPS utilities that can improve the quality of hard copy produced from screen captures is a procedure for switching colors. This capability is important because the grey scale rendering of colors on black-and-white printers may create poor contrasts if the original (default) color assignments are used. To access the CUPS utility for changing colors, the user need only choose **Configuration** under the **File** menu when the program is first initiated, or at any later time. Once you have chosen **Configuration**, to change colors you need to click the mouse on the **Change Colors** bar, and you will be presented with a 16 by 16 matrix of radio buttons that will allow you to change any color to any other color, or else to use predefined color switches, such as a color "reversal," or a conversion of all light colors to black, and all dark colors to white. (The screen captures given in this book were produced using the "reverse" color map.) Any such color changes must be redone when the program is restarted.

Other system parameters may likewise be set from the **File | Configuration** menu item. These include the path for temporary files that the program may create (or want to read), the mouse "double click" speed—important for those with slow reflexes—an added time delay to slow down programs on computers that are too fast, and a "check memory" option—primarily of interest to those making program modifications.

Those users wishing more information on the CUPS utilities should consult the CUPS Utilities Manual, written by Jaroslaw Tuszynski and William MacDonald, published by John Wiley and Sons. However, it is not necessary for casual users of CUPS programs to become familiar with the utilities. Such familiarity would only be important to someone wishing to write their own simulations using the utilities. The utilities are freely available for this purpose, for unrestricted noncommercial production and distribution of programs. However, users of the utilities who wish to write programs for commercial distribution should contact John Wiley and Sons.

1.5   Communicating With the Authors

Users of these programs should not expect that run-time errors will never occur! In most cases, such run-time errors may require only that the user restart the program; but in other cases, it may be necessary to reboot the computer, or even turn it off and on. The causes of such run-time errors are highly varied. In some cases, the program may be telling you something important about the physics or the numerical method. For example, you may be trying to use a numerical method beyond its range of applicability. Other types of run-time errors may have to do with memory or other limitations of your computer. Finally, although the programs in this book have been extensively tested, we cannot rule out the possibility that they may contain errors. (Please let us know if you find any! It would be most helpful if such problems were communicated by electronic mail, and with complete specificity as to the circumstances under which they arise.)

It would be best if you communicated such problems directly to the author of each program, and simultaneously to the editors of this book (the CUPS Directors), via electronic mail—see addresses listed below. Please feel free to communicate any suggestions about the programs and text which may lead to improvements in future editions. Since the programs have been provided in source code form, it will be possible for you to make corrections of any errors that you or we find in the future—provided that you send in the registration card at the back of the book, so that you can be notified. The fact that you have the source code will also allow you to make modifications and extensions of the programs. We can assume no responsibility for errors that arise in programs that you have modified. In fact, we strongly urge you to change the program name, and to add a documentary note at the beginning of the code of any modified programs that alerts other potential users of any such changes.

1.6 CUPS Courses and Developers

- **CUPS Directors**
  Maria Dworzecka, George Mason University (cups@gmu.edu)
  Robert Ehrlich, George Mason University (cups@gmu.edu)
  William MacDonald, University of Maryland (w_macdonald@umail.umd.edu)

- **Astrophysics**
  J. M. Anthony Danby, North Carolina State University (n38hs901@ncuvm. ncsu.edu)
  Richard Kouzes, Battelle Pacific Northwest Laboratory (rt_kouzes@pnl.gov)
  Charles Whitney, Harvard University (whitney@cfa.harvard.edu)

- **Classical Mechanics**
  Bruce Hawkins, Smith College (bhawkins@smith.bitnet)
  Randall Jones, Loyola College (rsj@loyvax.bitnet)

- **Electricity and Magnetism**
  Robert Ehrlich, George Mason University (rehrlich@gmuvax.gmu.edu)
  Lyle Roelofs, Haverford College (lroelofs@haverford.edu)
  Ronald Stoner, Bowling Green University (stoner@andy.bgsu.edu)
  Jaroslaw Tuszynski, George Mason University (cups@gmuvax.gmu.edu)

- **Modern Physics**
  Douglas Brandt, Eastern Illinois University (cfdeb@ux1.cts.eiu.edu)
  John Hiller, University of Minnesota, Duluth (jhiller@d.umn.edu)
  Michael Moloney, Rose Hulman Institute (moloney@nextwork.rose-hulman.edu)

- **Nuclear and Particle Physics**
  Roberta Bigelow, Willamette University (rbigelow@willamette.edu)
  John Philpott, Florida State University (philpott@fsunuc.physics.fsu.edu)
  Joseph Rothberg, University of Washington (rothberg@phys. washington.edu)

- **Quantum Mechanics**
  John Hiller, University of Minnesota Duluth (jhiller@d.umn.edu)
  Ian Johnston, University of Sydney (idj@suphys.physics.su.oz.au)
  Daniel Styer, Oberlin College (dstyer@physics.oberlin.edu)

- **Solid State Physics**
  Graham Keeler, University of Salford (g.j.keeler@physics.salford.ac.uk)
  Roger Rollins, Ohio University (rollins@chaos.phy.ohiou.edu)
  Steven Spicklemire, University of Indianapolis (steves@truevision.com)

- **Thermal and Statistical Physics**
  Harvey Gould, Clark University (hgould@vax.clarku.edu)
  Lynna Spornick, Johns Hopkins University
  Jan Tobochnik, Kalamazoo College (jant@kzoo.edu)

- **Waves and Optics**
  G. Andrew Antonelli, Wolfgang Christian, and Susan Fischer, Davidson College (wc@phyhost.davidson.edu)
  Robin Giles, Brandon University (giles@brandonu.ca)
  Brian James, Salford University (b.w.james@physics.salford.ac.uk)

1.7   Descriptions of all CUPS Programs

Each of the computer simulations in this book (as well as those in the eight other books comprised by the CUPS Project) are described below. The individual headings under which programs appear correspond to the nine CUPS courses. In several cases, programs are listed under more than one course. The number of programs listed under the Astrophysics, Modern Physics, and Thermal Physics courses is appreciably greater than the others, because several authors have opted to subdivide their programs into many smaller programs. Detailed inquiries regarding CUPS programs should be sent to the program authors.

ASTROPHYSICS PROGRAMS

STELLAR (Stellar Models), written by Richard Kouzes, is a simulation of the structure of a static star in hydrodynamic equilibrium. This provides a model of a zero age main sequence star, and helps the user understand the physical processes that exist in stars, including how density, temperature, and luminosity depend on mass. Stars are self-gravitating masses of hot gas supported by thermodynamic processes fueled by nuclear fusion at their core. The model integrates the four differential equations governing the physics of the star to reach an equilibrium condition which depends only on the star's mass and composition.

EVOLVE (Stellar Evolution), written by Richard Kouzes, builds on the physics of a static star, and considers (1) how a gas cloud collapses to become a main sequence star, and (2) how a star evolves from the main sequence to its final demise. The model is based on the same physics as the STELLAR program. Starting from a diffuse cloud of gas, a protostar forms as the cloud collapses and reaches a sufficient density for fusion to begin. Once a star reaches equilibrium, it remains for most of its life on the main sequence, evolving off after it has consumed its fuel. The final stages of the star's life are marked by rapid and dramatic evolution.

BINARIES is the driver program for all Binaries Programs (**VISUAL1, VISUAL2, ECLIPSE, SPECTRO, TIDAL, ROCHE, and ACCRDISK**).

VISUAL1 (Visual Binaries—Proper Motion), written by Anthony Danby, enables you to visualize the proper motion in the sky of the members of a visual binary system. You can enter the elements of the system and the mass ratio, as well as the speed at which the center of mass moves across the screen. The program also includes an animated three-dimensional demonstration of the elements.

VISUAL2 (Visual Binaries—True Orbit), written by Anthony Danby, enables you to select an apparent orbit for the secondary star with arbitrary eccentricity, with the primary at any interior point. The elements of the orbit are displayed. You can see the orbit animated in three dimensions, or can make up a set of "observations" based on the apparent orbit.

ECLIPSE (Eclipsing Binaries), written by Anthony Danby, shows simultaneously either the light curve and the orbital motion or the light curve and an animation of the eclipses. You can select the elements of the orbit and radii and magnitudes of the stars. A form of limb-darkening is also included as an option.

SPECTRO (Spectroscopic Binaries), written by Anthony Danby, allows you to select the orbital elements of a spectroscopic binary, and then shows simultaneously the velocity curve, the orbital motion, and a moving spectral line.

TIDAL (Tidal Distortion of a Binary), written by Anthony Danby, models the motion of a spherical secondary star around a primary that is tidally distorted by the secondary. You can select orbital elements, masses of the stars, a parameter describing the tidal lag, and the initial rate of rotation of the primary. The equations are integrated over a time interval that you specify. Then you can see the changes of the orbital elements, and the rotation of the primary, with time. You can follow the motion in detail around each revolution, or in a form where the equations have been averaged around each revolution.

ROCHE (The Photo-Gravitational Restricted Problem of Three Bodies), written by Anthony Danby, follows the two-dimensional motion of a particle that is subject to the gravitational attraction of two bodies in mutual circular orbits, and also, optionally, radiation pressure from these bodies. It is intended, in part, as background for the interpretation of the formation of accretion disks. Curves of zero velocity (that limit regions of possible motion) can be seen. The orbits can also be followed using Poincaré maps.

ACCRDISK (Formation of an Accretion Disk), written by Anthony Danby, follows some of the dynamical steps in this process. The dynamics is valid up to the initial formation of a hot spot, and qualititative afterward.

NBMENU is the driver program for all programs on the motion of N interacting bodies: **TWOGALAX, ASTROIDS, N-BODIES, PLANETS, PLAYBACK, and ELEMENTS**.

TWOGALAX (The Model of Wright and Toomres), written by Anthony Danby, is concerned with the interaction of two galaxies. Each consists of a central gravitationally attracting point, surrounded by rings of stars (which are attracted, but do not attract). Elements of the orbits of one galaxy relative to the other are selected, as is the initial distribution and population of the rings. The motion can be viewed as projected into the plane of the orbit of the galaxies, or simultaneously in that plane and perpendicular to it. The positions can be stored in a file for later viewing.

ASTROIDS (N-Body Application to the Asteroids), written by Anthony Danby, uses the same basic model, but a planet and a star take the place of the galaxies and the asteroids replace the stars. Emphasis is on asteroids all having the same period, with interest on periods having commensurability with the period of the planet. The orbital motion of the system can be followed. The positions can be stored in a file for later viewing. An asteroid can be selected, and the variation of its orbital elements can then be followed.

NBODIES (The Motion of N Attracting Bodies), written by Anthony Danby, allows you to choose the number of bodies (up to 20) and the total energy of the system. Initial conditions are chosen at random, consistent with this energy, and the resulting motion can be observed. During the motion various quantities, such as the kinetic energy, are displayed. The positions can be stored in a file for later viewing.

PLANETS (Make Your Own Solar System), written by Anthony Danby, is similar to the preceding program, but with the bodies interpreted as a star with planets. Initial conditions are specified through the choice of the initial elements of the planets. The positions can be stored in a file for later viewing.

PLAYBACK, written by Anthony Danby, enables a file stored by one of the preceding programs to be viewed.

ELEMENTS (Orbital Elements of a Planet), written by Anthony Danby, shows a three-dimensional animation that can be viewed from any angle.

GALAXIES is the driver program for Galactic Kinematics Programs: **ROTATION, OORTCONS, and ARMS21CM.**

ROTATION (The Rotation Curve of a Galaxy), written by Anthony Danby, first prompts you to "design" a galaxy, consisting of a central mass and up to five spheroids (that can be visible or invisible). It then displays the galaxy and can show the animated rotation or the rotation curve.

OORTCONS (Galactic Kinematics and Oort's Constants), written by Anthony Danby, allows you to design your galaxy, choose the location of the "sun" and a local region around it, and the to observe the kinematics in this region. It also shows graphs of radial velocity and proper motion in comparison with the linear approximation, and computes the Oort constants.

ARMS21CM (The Spiral Structure of a Galaxy), written by Anthony Danby, allows you to design your galaxy, construct a set of spiral arms, and select the position of the "sun." Then, for different galactic longitudes, you can see observed profiles of 21 cm lines.

ATMOS (Stellar Atmospheres), written by Charles Whitney, permits the user to select a constellation, see it mapped on the computer screen, point to a star, and see it plotted on a brightness-color diagram. The user's task is to build a model atmosphere that imitates the photometric properties of observed stars. This is done by specifying numerical values for three basic stellar parameters: radius, mass, and luminosity. The program then builds the model and displays it on the brightness-color diagram, and it also plots the spectrum and the detailed thermodynamic structure of the atmosphere. With this program the user may investigate the relation between stellar parameters and the thermal properties of the gas in the atmosphere. Two atmospheres may be superposed on the graphs, for easier comparison.

PULSE (Stellar Pulsations), written by Charles Whitney, illustrates stellar pulsation by simulating the thermo-mechanical behavior of a "star" modelled by a self-gravitating gas divided by spherical elastic shells. The elastic shells resemble a set of coupled oscillators. The program solves for the modes of small-amplitude motion, and it uses Fourier synthesis to construct motions for arbitrary starting conditions. The screen displays the thermodynamic structure and surface properties, such as temperature, pressure, and velocity. Animation displays the nature of the pulsation. By showing the motions, temperatures, and energy flux, the program demonstrates the heat engine acting inside the pulsating star. The motions of the shells and the spatial Fourier decomposition into eigenmodes are displayed simultaneously, and this will help you visualize the meaning of the Fourier components.

CLASSICAL MECHANICS PROGRAMS

GENMOT (The Motion Generator), written by Randall Jones, allows you to solve numerically any differential equation of motion for a system with up to three degrees of freedom and display the time evolution of the system in a wide variety of formats. Any of the dynamical variables or any function of those variables may be displayed graphically and/or numerically and a wide range of animations may be constructed. Since the Motion Generator can be used to solve any second-order differential equation, it can also be used to study systems analyzed by Lagrangian methods. Real world coordinates may be constructed as functions of generalized coordinates so that simulations of the actual system can be constructed.

ROTATE (Rotation of 3-D Objects), written by Randall Jones, is designed to aid in the visualization of the dynamical variables of rotational motion. It will allow you to observe the 3-D motion of rotating objects in a controlled fashion, running the simulation faster, slower, or in reverse while displaying the corresponding evolution of the angular velocity, the angular momentum and the torque. It will display the motion from the fixed frame and from the body frame to help in understanding the translation between these two descriptions of the motion. By using the stereographic feature of the program you can create a genuine 3-D representation of the motion of the quantities.

COUPOSC (Coupled Oscillators), written by Randall Jones, is designed to investigate a wide range of harmonic systems. Given a set of objects and springs connected in one or two dimensions, the simulation can solve the problem by generating the normal mode frequencies and their corresponding motions. It can take any set of initial conditions and resolve them into their component normal mode motions or take any set of initial mode occupations and display the corresponding motions of the objects. It can also determine the motion of the system when it is acted on by external forces. In this case the total forces are no longer harmonic, so the solution is generated numerically. The harmonic analysis, however, still provides an important tool for investigating and understanding the subsequent motion.

ANHARM (Anharmonic Oscillators), written by Bruce Hawkins, simulates oscillations of various types: pendulum, simple harmonic oscillator, asymmetric, cubic, Vanderpol, and a mass in the center of a spring with fixed ends. Nonlinear behavior is emphasized. The user may choose to view one to four graphs of the motion simultaneously, along with the potential diagram and a picture of the moving object. Graphs that may be viewed are x vs. t, v vs. t, v vs. x, the Poincaré diagram, and the return map. Tools are provided to explore parameter space for regions of interest. Fourier analysis is available, resonance diagrams can be plotted, and the period can be plotted as a function of energy. Includes a tutorial demonstrating the usefulness of phase plots and Poincaré plots.

ORBITER (Gravitational Orbits), written by Bruce Hawkins, simulates the motion of up to five objects with mutually gravitational attraction, and any reasonable number of additional objects moving in the gravitation field of the first five. The motion may be viewed in up to six windows simultaneously: windows centered on a particular body, on the center of mass, stationary in the universe frame, or rotating with the line joining the two most massive bodies. A menu of available system includes the solar system, the sun/earth/moon system; the sun, Jupiter, and its moons; the sun, earth, and Saturn, demonstrating retrograde motion; the sun, Jupiter, and a comet; and a pair of binary stars with a comet. Bodies my be added to any system, or a new system created using either numerical coordinates or the mouse. Bodies may be replicated to demonstrate the sensitivity of orbits to initial conditions.

COLISION (Collisions), written by Bruce Hawkins, simulates two-body collisions under any of a number of force laws: Coulomb with and without shielding and truncation, hard sphere, soft sphere (harmonic), Yukawa, and Woods-Saxon. Collision may be viewed in the laboratory and center of mass systems simultaneously, with or without momentum diagrams. Includes a tutorial on the usefulness of the center of mass system, one on the kinematics of relativistic collisions, and one on cross section. Plots cross section against scattering parameter, and compares collisions at different parameters.

ELECTRICITY AND MAGNETISM PROGRAMS

FIELDS (Analysis of Vector and Scalar Fields), written by Jarek Tuszynski, displays scalar and vector fields for any algebraic or trigonometric expression entered by the user. It also computes numerically the divergence, curl, and Laplacian for the vector fields, and the gradient and Laplacian for the scalar fields. Simultaneous displays of selected quantities are provided in user-selected planes, using vector, contour, or 3-D plots. The program also allows the user to define paths along which line integrals are computed.

GAUSS (Gauss' Law), written by Jarek Tuszynski, treats continuous charge distributions having spherical or cylindrical symmetry, and those that vary as a function of the x-coordinate only. The program allows the user to enter an arbitrary function to define either the electric field magnitude, the potential, or the charge density. It then computes the other two functions by numerical differentiation or integration, and displays all three functions. Finally, the program allows the user to enter a "comparison function," which is plotted on the same graph, so as to check whether his analytic solutions are correct.

POISSON (Poisson's Equation Solved on a Grid), written by Jarek Tuszynski, solves Poisson's equation iteratively on a 2-D grid using the method of simultaneous over-relaxation. The user can draw arbitrary systems consisting of line charges, and charged conducting cylinders, plates, and wires, all infinite in extent perpendicular to the grid. After iteratively solving Poisson's equation, the program displays the results for the potential, electric field, or the charge density (found from the Laplacian of the potential), in the form of contour, vector, or 3-D plots. In addition, many other program features are available, including the ability to specify surfaces, along which the potential varies according to some algebraic function specified by the user.

IMAG&MUL (Image Charges and Multipole Expansion), written by Lyle Roelofs and Nathaniel Johnson, allows users to explore two approaches to the solution of Laplace's equation—the image charge method and expansion in multipole moments. In the image charge mode (IC) the user is presented with a variety of configurations involving conducting planes and point charges and is asked to "solve" each by placing image charges in the appropriate locations. The program displays the electric field due to all point charges, real and image, and a solution can be regarded as successful with the field due to all charges is everywhere orthogonal to all conducting surfaces. Solutions can then be examined with a variety of included software "tools." The multipole expansion (ME) mode of the program also permits a "hands-on" exploration of standard electrostatic problems, in this case the "exterior" problem, i.e., the determination of the field outside a specified equipotential surface. The program presents the user with a variety of azimuthally symmetric equipotential surfaces. The user "solves" for the full potential by adding chosen amounts of the (first six) multipole moments. The screen shows the contours of the summed potential and the problem is "solved" when the innermost contour matches the given equipotential surface as closely as possible.

ATOMPOL (Atomic Polarization), written by Lyle Roelofs and Nathaniel Johnson, is an exploration of the phenomenon of atomic polarization. Up to 36 atoms of controllable polarizibility are immersed in an external electric field. The program solves for and displays the field throughout the region in which the atoms are located. A closeup window shows the polarization of selected atoms and software "tools" allow for further analysis of the resulting electric fields. Use of this program improves the student's understanding of polarization, the interaction of polarized entities and the atomic origin of macroscopic polarization, the latter via study of closely spaced clusters of polarizable atoms.

DIELECT (Dielectric Materials), written by Lyle Roelofs and Nathaniel Johnson, is a simulation of the behavior of linear dielectric materials using a cell-based approach. The user controls either the polarization or the susceptibility of each cell in a (25×25) grid (with assume uniformity in the third direction). Full self-consistent solutions are obtained via an iterative relaxation method and the fields P, E, or D are displayed. The student can investigate the self-interactions of polarized materials and many geometrical effects. Use of this program aids the student in developing understanding of the subtle relations among and meaning of P, E, and D.

ACCELQ (Fields From an Accelerated Charge), written by Ronald Stoner, simulates the electromagnetic fields in the plane of motion generated by a point charge that is moving and accelerating in two dimensions. The user chooses from among seven predefined trajectories, and sets the values of maximum speed and viewing time. The electric field pattern is recomputed after each change of trajectory or parameter; thereafter, the user can investigate the electric field, magnetic field, retarded potentials, and Poynting-vector field by using the mouse as a field probe, by using gridded overlays, or by generating plots of the various fields along cuts through the viewing plane.

QANIMATE (Fields From an Accelerated Charge—Animated Version), written by Ronald Stoner, is an interactive animation of the changing electric field pattern generated by a point electric charge moving in two dimensions. Charge motion can be manipulated by the user from the keyboard. The display can include electric field lines, radiation wave fronts, and their points of intersection. The motion of the charge is controlled by the using **arrow** keys to accelerate and steer much like the accelerator and steering wheel of a car, except that acceleration must be changed in increments, and the **Space** bar can used to engage or disengage the steering. With steering engaged, the charge will move in a circle. Unless the acceleration is made zero, the speed will increase (or decrease) to the maximum (minimum) possible value. At constant speed and turning rate, the charge can be controlled by the **Space** bar alone.

EMWAVE (Electromagnetic Waves), written by Ronald Stoner, uses animation to illustrate the behavior of electric and magnetic fields in a polarized plane electromagnetic wave. The user can choose to observe the wave in free space, or to see the effect on the wave of incidence on a material interface, or to see the effects of optical elements that change its polarization. The user can change the polarization state of the incident wave by specifying its Stokes parameters. Standing electromagnetic waves can be simulated by combining the incident travelling wave with a reflected wave of the same amplitude. The user can do that by choosing appropriate values of the physical properties of the medium on which the incident wave impinges in one of the animations.

MAGSTAT (Magnetostatics), written by Ronald Stoner, computes and displays magnetic fields in and near magnetized materials. The materials are uniform and have 3-D shapes that are solids of revolution about a vertical axis. The shape of the material can be modified or chosen from a data input screen. The user has the option of generating the fields produced by a permanently and uniformly magnetized object, or of generating the fields of a magnetizable object placed in an otherwise uniform external field. Besides choosing the shape and aspect ratio of the object, the user can vary the magnetic permeability of the magnetizable material, and choose among three fields to display: magnetic induction (B), magnetic field strength (H), and magnetization (M). Each of these fields can be displayed or explored in several different ways. The algorithm for computing the fields uses a superposition of Chebyschev polynomial approximants to the H field due to "rings" of "magnetic charge."

MODERN PHYSICS PROGRAMS

NUCLEAR (Nuclear Energetics and Nuclear Counting), written by Michael Moloney, deals with basic nuclear properties related to mass, charge, and energy, for approximately 1900 nuclides. Graphs are available involving binding energy, mass, and Q values of a variety of nuclear reactions, including alpha and beta decays. Part 2 deals with simulating the statistics of counting with a Geiger-Muller tube. This part also simulates neutron activation, and the counting behavior as neutron flux is turned on and off. Finally, a decay chain from A to B to C is simulated, where half-lives may be changed, and populations are graphed as a function of time.

GERMER (Davisson-Germer and G. P. Thomson Experiments), written by Michael Moloney, simulates both the Davisson-Germer and G. P. Thomson experiments with electrons scattering from crystalline materials. Stress is laid on the behavior of electrons as waves; similarities are noted with scattering of x-rays. The exercises encourage students to understand why peaks and valleys in scattered electrons occur where they do.

QUANTUM (one-dimensional Quantum Mechanics), written by Douglas Brandt, is a program that has four sections. The first section allows users to investigate the uncertainty principle for specified wavefunctions in position or momentum space. The second section allows users to investigate the time evolution of wavepackets under various dispersion relations. The third section allows users to investigate solutions to Schrödinger's equation for asymptotically free solutions. The user can input a barrier and the program calculates reflection and transmission coefficients for a range of energies and show wavepacket time evolution for the barrier potential. The fourth section is similar to the third, except that it allows the user to investigate bound solutions to Schrödinger's equation. The program calculates the bound state Hamiltonian eigenvalues and spatial eigenfunctions.

RUTHERFD (Rutherford Scattering), written by Douglas Brandt, is a program for investigating classical scattering of particles. A scattering potential can be chosen from a list of predefined potentials or an arbitrary potential can be input by the user. The computer generates scattering events by randomly picking impact parameters from a distribution defined by beam parameters specified by the user. It displays the results of the scattering on a polar histogram and on a detailed histogram to help users gain insight into differential scattering cross section. A scintillation mode can be chosen for users that want more appreciation of the actual experiments of Geiger and Marsden. A "guess the scatterer" mode is available for trying to gain appreciation of how scattering experiments are used to infer properties of the scatterers.

SPECREL (Special Relativity), written by Douglas Brandt, is a program to investigate special relativity. The first section is to investigate change of coordinate systems through Minkowski diagrams. The user can define coordinates of objects in one reference frame and the computer calculates the coordinates in a user-selectable coordinate system and displays the objects in both reference frames. The second section allows users to view clocks that are in relative motion. A clock can be given an arbitrary trajectory through space-time and the readings of various clocks can be viewed as the clock follows that trajectory. A third section allows users to observe collisions in different reference frames that are related by Lorentz transformations or by Gallilean transformations.

LASER (Lasers), written by Michael Moloney, simulates a three-level laser, with the user in control of energy level parameters, temperature, pump power, and end mirror transmission. Atomic populations may be graphically tracked from thermal equilibrium through the lasing threshold. A mirror cavity simulation is available which uses ray tracing. This permits study of cavity stability as a function of mirror shape and position, as well as beam shape characteristics within the cavity.

HATOM (Hydrogenic Atoms), written by John Hiller, computes eigenfunctions and eigenenergies for hydrogen, hydrogenic atoms, and single-electron diatomic ions. Hydrogenic atoms may be exposed to uniform electric and magnetic fields. Spin interactions are not included. The magnetic interaction used is the quadratic Zeeman term; in the absence of spin-orbit coupling, the linear term adds only a trivial energy shift. The unperturbed hydrogenic eigenfunctions are computed directly from the known solutions. When external fields are included, approximate results are obtained from basis-function expansions or from Lanczos diagonalization. In the diatomic case, an effective nuclear potential is recorded for use in calculation of the nuclear binding energy.

NUCLEAR AND PARTICLE PHYSICS PROGRAMS

NUCLEAR (Nuclear Energetics and Counting), written by Michael Moloney, is included here, but is described under the Modern Physics Heating.

SHELLMOD (Nuclear Models), written by Roberta Bigelow, calculates energy levels for spherical and deformed nuclei using the single particle shell model. You can explore how the nuclear potential shape, the spin-orbit interaction, and deformation affect both the order and spacing of nuclear energy levels. In addition, you will learn how to predict spin and parity for single particle states.

NUCRAD (Interaction of Radiation With Matter), written by Roberta Bigelow, is a simulation of alpha particles, muons, electrons, or photons interacting with matter. You will develop an understanding of how ranges, energy losses, and random particle paths depend on materials, radiation, and incident energy. As a specific application, you can explore photon and electron interactions in a sodium iodide crystal which determines the energy response of a radiation detector.

ELSCATT (Electron-Nucleus Scattering), by John Philpott, is an interactive software tool that demonstrates various aspects of electron scattering from nuclei. Specific features include the relativistic kinematics of electron scattering, densities and form factors for elastic and inelastic scattering, and the nuclear Coulomb response. The simulation illustrates how detailed nuclear structure information can be obtained from electron scattering measurements.

TWOBODY (Two-Nucleon Interactions), by John Philpott, is an interactive software tool that illuminates many features of the two-nucleon problem. Bound state wavefunctions and properties can be calculated for a variety of interactions that may include non-central parts. Phase shifts and cross sections for pp, pn, and nn scattering can be calculated and compared with those obtained experimentally. Spin-polarization features of the cross sections can be extensively investigated. The simulation demonstrate the richness of the two-nucleon data and its relation to the underlying nucleon-nucleon interaction.

RELKIN (Relativistic Kinematics), by Joseph Rothberg, is an interactive program to permit you to explore the relativistic kinematics of scattering reactions and two-body particle decays. You may choose from among a large number of initial and final states. The initial momentum of the beam particle and the center of mass angle of a secondary can also be specified. The program displays the final state vector momenta in both the lab system and center of the mass system along with numerical values of the most important kinematic quantities. The program may be run in a Monte Carlo mode, displaying a scatter plot and histogram of selected variables. The particle data base may be modified by the user and additional reactions and decay modes may be added.

DETSIM (Particle Detector Simulation), by Joseph Rothberg, is an interactive tool to allow you to explore methods of determining parameters of a decaying particle or scattering reaction. The program simulates the response of high-energy particle detectors to the final-state particles from scattering or decays. The detector size and location may be specified by the user as well as its energy and spatial resolution. If the program is run in a Monte Carlo mode, detector hit information for each event is written to a file. This file can be read by a small reconstruction and plotting program. You can easily modify one of the example reconstruction programs that are provided to determine the mass, momentum, and other properties of the initial particle or state.

QUANTUM MECHANICS PROGRAMS

BOUND1D (Bound States in One Dimension), written by Ian Johnston, is a tool which allows you to explore energy eigenfunctions for an electron in various potential wells, which can be square, parabolic, ramped, asymmetric, double or Coulombic. The first part of the program deals with finding the eigenvalues and eigenfunctions of different wells. You may find them yourself, using a "hunt and shoot" method, or else the program will compute the eigenvalues automatically, by counting the number of nodes to determine where the eigenvalues occur. The second part of the program looks at properties of eigenfunctions normalization, orthogonality and the evaluation of many kinds of overlap integrals. The third part examines time development of general states made up of a superposition of bound state eigenfunctions. Facility is provided for you to incorporate your own procedures to specify different potential wells or different overlap integrals.

SCATTR1D (Scattering in One Dimension), written by John Hiller, solves the time-independent Schrödinger equation for stationary scattering states in one-dimensional potentials. The wavefunction is displayed in a variety of ways, and the transmission and reflection probabilities are computed. The probabilities may be displayed as functions of energy. The computations are done by numerically integrating the Schrödinger equation from the region of the transmitted wave, where the wavefunction is known up to some overall normalization and phase, to the region of the incident wave. There the reflected and incident waves are separated. The potential is assumed to be zero in the incident region and constant in the transmitted region.

QMTIME (Quantum Mechanical Time Development), written by Daniel Styer, simulates quantal time development in one dimension. A variety of initial wave packets (Gaussian, Lorentzian, etc.) can evolve in time under the influence of a variety of potential energy functions (step, ramp, square well, harmonic oscillator, etc.) with or without an external driving force. A novel visualization technique simultaneously displays the magnitude and phase of complex-valued wave functions. Either position-space or momentum-space wave functions, or both, can be shown. The program is particularly effective in demonstrating the classical limit of quantum mechanics.

LATCE1D (Wavefunctions on a one-dimensional Lattice), written by Ian Johnston, is a tool which allows you to explore energy eigenfunctions for an electron in a lattice made up of a number of simple potential wells (up to twelve), which can be square, parabolic, or Coulombic. You may find the eigenvalues yourself, using a "hunt and shoot" method, or allow the program to compute them automatically. You can firstly explore regular lattices, where all wells are the same and spaced at regular intervals. These will demonstrate many of the properties of regular crystals, particularly the existence of energy bands. Secondly you can change the width, depth or spacing of any of the wells, which will mimic the effect of impurities or other irregularities in a crystal. Lastly you can apply an external electric across the lattice. Facility is provided for you to incorporate your own procedures to calculate wells, lattice arrangements or external fields of their own choosing.

BOUND3D (Bound States in Three Dimensions), written by Ian Johnston, is a tool which allows you to explore energy eigenfunctions for an particle in a spherically symmetric potential well, which can be square, parabolic, Coulombic, or several other shapes of importance in molecular or nuclear applications. The first part of the program deals with finding the eigenvalues and eigenfunctions of different wells, assuming that the angular part of the wavefunctions are spherical harmonics. You may find them yourself for a given angular momentum quantum number using a "hunt and shoot" method, or else the program will compute the eigenvalues automatically, by counting the number of nodes to determine where the eigenvalues occur. The second part of the program looks at properties of eigenfunctions normalization, orthogonality and the evaluation of many kinds of overlap integrals. Facility is provided for you to incorporate your own procedures to specify different potential wells or different overlap integrals.

IDENT (Identical Particles in Quantum Mechanics), written by Daniel Styer, shows the probability density associated with the symmetrized, antisymmetrized, or nonsymmetrized wave functions of two noninteracting particles moving in a one-dimensional infinite square well. It is particularly valuable for demonstrating the effective interaction of noninteracting identical particles due to interchange symmetry requirements.

SCATTR3D (Scattering in Three Dimensions), written by John Hiller, performs a partial-wave analysis of scattering from a spherically symmetric potential. Radial and 3-D wave functions are displayed, as are phase shifts, and differential and total cross sections. The analysis employs an expansion in the natural angular momentum basis for the scattering wavefunction. The radial wavefunctions are computed numerically; outside the region where the potential is important they reduce to a linear combination of Bessel functions which asymptotically differs from the free radial wavefunction by only a phase. Knowledge of these phase shifts for the dominant values of angular momentum is used to approximate the cross sections.

CYLSYM (Cyllindrically Symmetric Potentials), written by John Hiller, solves the time-independent Schrödinger equation Hu=Eu in the case of a cylindrically symmetric potential for the lowest state of a chosen parity and magnetic quantum number. The method of solution is based on evolution in imaginary time, which converges to the state of the lowest energy that has the symmetry of the initial guess. The Alternating Direction Implicit method is used to solve a diffusion equation given by $HU = -\hbar \partial U / \partial t$, where H is the Hamiltonian that appears in the Schrödinger equation. At large times, U is nearly proportional to the lowest eigenfunction of H, and the expectation value $\langle H \rangle = \langle U | H | U \rangle / \langle U | U \rangle$ is an estimate for the associated eigenenergy.

SOLID STATE PHYSICS

LATCE1D (Wavefunctions for a one-dimensional Lattice), written by Ian Johnston, and included here, is described under the Quantum Mechanics heading.

SOLIDLAB (Build Your Own Solid State Devices), written by Steven Spicklemire, is a simulation of a semiconductor device. The device can be "drawn" by the user, and the characteristics of the device adjusted by the user during the simulation. The user can see how charge density, current density, and electric potential vary throughout the device during its operation.

LCAOWORK (Wavefunctions in the LCAO Approximation), written by Steven Spicklemire, is a simulation of the interaction of 2-D atoms within small atomic clusters. The atoms can be adjusted and moved around while their quantum mechanical wavefunctions are calculated in real time. The student can investigatge the dependence of various properties of these atomic clusters on the properties of individual atoms, and the geometric arrangement of the atoms within the cluster.

PHONON (Phonons and Density of States), written by Graham Keeler, calculates and displays phonon dispersion curves and the density of states for a number of different 3-D cubic crystal structures. The displays of the dispersion curves show realistic curves and allow the user to study the effect of changing the interatomic forces between nearest and further neighbor atoms and, for diatomic crystal structures, changing the ratio of the atomic masses. The density of states calculation shows how the complex shapes of real densities of states are built up from simpler distributions for each mode of polarization, and enables the user to match the features of the distribution to corresponding features on the dispersion curves. In order to help with visualization of the crystal lattices involved, the program also shows 3-D projections of the different crystal structures.

SPHEAT (Calculations of Specific Heat), written by Graham Keeler, calculates and displays the temperature variation of the lattice specific heat for a number of different theoretical models, including the Einstein model and the Debye model. It also makes the calculation for a computer simulation of a realistic density of states, in which the user can vary the important parameters of the crystal, including those affecting the density of state. The program can display the results for a small region near the origin, and as a T-cubed plot to enable the user to investigae the low temperature limit of the specific heat, or in the form of the equivalent Debye temperature to enhance a study of the deviations from the Debye model. The Schottky specific heat anomaly can also be investigated.

BANDS (Energy Bands), written by Roger Rollins, calculates and displays, for easy comparison, the energy dispersion curves and corresponding wavefunctions for an electron in a 1-D symmetric $V(x) = V(-x)$ periodic potential of arbitrary shape and of strength V_0. The method used is based on an exact, non-perturbative approach so that the energy dispersion curves and band gaps can be obtained for large V_0. Wavefunctions can be displayed, and compared with one another, by clicking the mouse on the desired states on the energy dispersion curve. Changes in band strtucture can be followed as changes are made in the shape of the potential. The variation of the band gaps with V_0 is calculated and compared with the two opposite limits of very weak V_0 (perturbation method) and very strong V_0 (isolated atom). Even the experienced condensed matter researcher may be surprised by some of the results! Open-ended class discussions can result from the interesting physics found in these conceptually simple model calculations.

PACKET (Electron Wavepacket in a 1-D Lattice), written by Roger Rollins, shows a live animation, calculated in real time, demonstrating how an electron wavepacket in a metal or semiconducting crystal moves under the influence of external forces. The time-dependent Schrödinger equation is solved in a tight binding approximation, including the external force terms, and the motion of the wavepacket is obtained directly. The main objective of the simulation is to show that an electron wavepacket formed from states with energies near the top of an energy band is accelerated in a direction *opposite* to the direction of the external force; it has a *negative* effective mass! The simulation deals with motion in a 1-D lattice but the concepts are applicable to the full 3-D motion of an electron in a real crystal. Numerical experiments on the motion of the packet explore interesting physics questions such as: how does constant applied force affect the periodic motion of a packet? when does the usual semiclassical model fail? what happens to the dynamics of the packet when placed in a superlattice with lattice constant twice that of the original lattice?

THERMAL AND STATISTICAL PHYSICS PROGRAMS

ENGDRV, written by Lynna Spornick, is a driver program for **ENGINE, DIESEL, OTTO, and WANKEL**. These programs provide an introduction to the thermodynamics of engines.

ENGINE (Design Your Own Engine), written by Lynna Spornick, lets the user design an engine by specifying the processes (adiabatic, isobaric, isochoric [constant volume], and isothermic) in the engine's cycle, the engine type (reversible or irreversible), and the gas type (helium, argon, nitrogen, or steam). The thermodynamic properties (heat exchanged, work done, and change in internal energy) for each process and the engine's efficiency are computed.

DIESEL, OTTO, and WANKEL, written by Lynna Spornick, provide animations of each of these types of engine. Plots of the temperature versus entropy and the pressure versus volume for the cycles are show with the engine's current thermodynamic conditions indicated.

PROBDRV, written by Lynna Spornick, is a driver program for **GALTON, POISEXP, TWOD, KAC, and STADIUM**. Subprograms GALTON, POISEXP, and TWOD provide an introduction to probability and subprograms KAC and STADIUM provide an introduction to statistics.

GALTON (A Galton Board), written by Lynna Spornick, models either a traditional Galton Board or a customized Galton Board with traps, reflecting, and/or absorbing walls. GALTON demonstrates the binominal and normal distributions, the laws of probability, and the central limit theorem.

POISEXP (Poisson Probability Distribution in Nuclear Decay), written by Lynna Spornick, uses the decay of radioactive atoms to describe the Poisson and the exponential distributions.

TWOD (2-D Random Walk), written by Lynna Spornick, models a random walk in two dimensions. A "drunk," taking equal-length steps, is required to walk either on a grid or on a plane. TWOD demonstrates the joint probability of two independent processes, the binominal distribution, and the Rayleigh distribution.

KAC (A Kac Ring), written by Lynna Spornick, uses a Kac ring to demonstrate that large mechanical systems, whose equations of motion are solvable and which obey time reversal and have a Poincaré cycle, can also be described by statistical models.

STADIUM (The Stadium Model), written by Lynna Spornick, uses a stadium model to demonstrate that there exists mechanical systems whose equations of motion are solvable but whose motion is not predictable because of the system's chaotic nature.

ISING (Ising Model in One and Two Dimensions), written by Harvey Gould, allows the user to explore the static and dynamic properties of the 1- and 2-D Ising model using four different Monte Carlo algorithms and three different ensembles. The choice of the Metropolis algorithm allows the user to study the Ising model at constant temperature and external magnetic field. The orientation of the spins is shown on the screen as well as the evolution of the total energy or magnetization. The mean energy, magnetization, heat capacity, and susceptibility are monitored as a function of the number of configurations that are sampled. Other computed quantities include the equilibrium-averaged energy and magnetization autocorrelation functions and the energy histogram. Important physical concepts that can be studied with the aid of the program include the Boltzmann probability, the qualitative behavior of systems near critical points, critical exponents, the renormalization group, and critical slowing down. Other algorithms that can be chosen by the user correspond to spin exchange dynamics (constant magnetization), constant energy (the demon algorithm), and single cluster Wolff dynamics. The latter is particularly useful for generating equilibrium configurations at the critical point.

MANYPART (Many Particle Molecular Dynamics), written by Harvey Gould, allows the user to simulate a dense gas, liquid, or solid in two dimensions using either molecular dynamics (constant energy, constant volume) or Monte Carlo (constant temperature, constant volume) methods. Both hard disks and the Lennard-Jones interaction can be chosen. The trajectories of the particles are shown as the system evolves. Physical quantities of interest that are monitored include the pressure, temperature, heat capacity, mean square displacement, distribution of the speeds and velocities, and the pair correlation function. Important physical concepts that can be studied with the aid of the program include the Maxwell-Boltzmann probability distribution, fluctuations, equation of state, correlations, and the importance of chaotic mixing.

FLUIDS (Thermodynamics of Fluids), written by Jan Tobochnik, allows the user to explore the fluid (gas and liquid) phase diagrams for the van der Waals model and water. The user chooses four phase diagrams from among the following choices: PT, Pv, vT, uT, sT, uv, and sv, where P is the pressure, T is the temperature, v is the specific volume, S is the specific entropy, and u is the specific internal energy. The program reads in the coexistence table for the van der Waals model and water, and uses it along with an empirical formula for the water free energy and the free energy derived from the van der Waals model. Given v and u, any thermodynamic quantity can be calculated. For the van der Waals model thermodynamic quantities also can be calculated from the other thermodynamic state variables. The user can draw a straight line path in one phase diagram and see how this path looks in the other phase diagrams. The user also can extract all important thermodynamic data at any point in a phase diagram.

QMGAS1 (Quantum Mechanical Gas—Part 1), written by Jan Tobochnik, does the numerical calculations necessary to solve for the thermodynamic properties of quantum ideal gases, including photons in blackbody radiation, ideal bosons, phonons in the Debye theory, non-interacting fermions, and the classical limits of these systems. The user chooses the type of statistics (Bose-Einstein, Fermi-Dirac, or Maxwell-Boltzmann), the dimension of space, the form of the dispersion relation (restricted to simple powers), whether or not the particles have a non-zero chemical potential, and whether or not there is a Debye cutoff. The program then allows the user to build up a table of thermodynamic data, including the energy, specific heat, and chemical potential as a function of temperature. This data and various distribution functions and the density of states can be plotted.

QMGAS2 (Quantum Mechanical Gas—Part2), written by Jan Tobochnik, implements a Monte Carlo simulation of a finite number of quantum particles fluctuating between various states in a finite k-space (k is the wavevector). The program orders the possible energy states into an energy level diagram and then allows particles to move from one state to another according to the usual Boltzman probability distribution. Bosons are restricted so that they may not pass through each other on the energy level diagram; fermions are further restricted so that no two fermions may be in the same state; classical particles have no restrictions. In this way indistinguishability is correctly implemented for bosons and fermions. The user chooses the type of particle, the number of particles, the size and dimension of k-space, and the temperature. During the simulation the user sees a representation of the state occupancy and plots of the average energy, the instantaneous energy, and the distribution of energy among the states, also shown are results for the average energy, specific heat, and the occupancy of the ground state.

WAVES AND OPTICS PROGRAMS

DIFFRACT (Interference and Diffraction), by Robin Giles, simulates some of the fundamental wave behaviors in Fresnel and Fraunhofer Diffraction, and other Interference and Coherence effects. In particular you will be able to study diffraction phenomena associated with a point or a set of points and a slit or set of slits using the Huyghens construction. You can also use a method developed by Cornu—the Cornu Spiral—to examine diffraction from one or two slits or one or two obstacles. You can study Fresnel and Fraunhofer diffraction with a single slit or set of slits, a rectangular aperture and a circular aperture. Finally you can study Partial Coherence and fringe visibility in interference and diffraction observations. In the latter example you will be able to study the Michelson Stellar Interferometer and measure the separation distance in a double star and measure the diameter of single stars.

SPECTRUM (Applications of Interfence and Diffraction), by Robin Giles, simulates the uses and modes of operation of four important optical instruments—the Diffraction Grating, the Prism Spectrometer, the Michelson Interferometer and the Fabry-Perot Interferometer. You will look at the nature of the spectra, simulated interference patterns, and the question of resolving power.

WAVE (One-Dimensional Waves), by Wolfgang Christian, Andrew Antonelli, and Susan Fischer, uses finite difference methods to study the time evolution of the following partial differential equations: classical wave, Schrödinger, diffusion, Klein-Gordon, sine-Gordon, phi four, and double sine-Gordon. The user may vary the initial function and boundary conditions. Unique features of the program include mouse-driven drawing tools that enable the user to create sources, segments, and detectors anywhere inside the medium. Double-clicking on a segment allows the user to edit properties such as index of refraction or potential in order to model barrier problems such as thin film interference filters or the Ramsauer-Townsend effect in optics and quantum mechanics, respectively. Various types of analysis can be performed, including detector value, space-time, Fourier analysis and energy density.

CHAIN (One-Dimensional Lattice of Coupled Oscillators), by Wolfgang Christian, Andrew Antonelli, and Susan Fischer, allows the user to examine the time evolution of a 1-D lattice of coupled oscillators. These oscillators are represented on screen as a chain of masses undergoing vertical displacement. The program allows the user to examine how the application of Newtonian mechanics to these masses leads to traveling and standing waves. The relationship between the lattice spacing and other properties such as dispersion, band gaps, and cut-off frequency can be examined. Each mass can be assigned linear, quadratic, and cubic nearest neighbor interactions as well as a time-dependent external force function. Global properties such as the total energy in the lattice or the Fourier transform of the lattice can be displayed as well as the time evolution of a single mass's dynamical variables.

FOURIER (Fourier Analysis and Synthesis), written by Brian James, allows investigation of Fourier analysis and 1-D and 2-D Fourier transforms. In Fourier analysis users can choose from several predefined functions or enter their own functions either algebraically, numerically or graphically. The build-up of a periodic function is illustrated as successive terms of the Fourier series are added in, and the effects of dispersion and attenuation on the evolution of the synthesized waveform can then be investigated. One- and two-dimensional discrete Fourier transforms can be produced for a range of standard and user-entered functions. The effects of filters on the inverse transforms are illustrated. The 2-D transforms are shown as surface and contour plots. Image processing can be illustrated by filtering the transforms of gray level images so that when the inverse transforms are displayed it can be seen that the images have been modified.

RAYTRACE (Ray Tracing and Lenses), by Brian James, lets the user explore the applications of ray tracing in geometrical optics. The fundamental principle of Fermat can be illustrated by plotting the path of a ray through two different materials between fixed points. The variation of the path of a ray through a region of changing refractive index can be used to investigate the formation of mirages. The variation of pulse delay in a fiber can be calculated as a function of its parameters and the characteristics of optical communication fibers are considered. The formation of primary and secondary rainbows due to dispersion of refractive index can be displayed. The matrix method of tracing rays through lenses can be used to investigate the images formed and show how aberrations in images arise and may be reduced.

QUICKRAY (Quick Ray Tracing), by John Philpott, can be used to demonstrate ray diagrams for a single thin lens or spherical mirror. The object and image are shown, along with the three principal rays that proceed from the object towards the observer. You can use the mouse to move the object, the position of the lens or mirror or to change the focal length of the lens or mirror. The principal rays are continuously redrawn while any of these adjustments are made. The simulation handles converging and diverging lenses and concave and convex mirrors. Thus students can quickly get an intuitive feel for real and virtual image formation under a variety of circumstances.

Acknowledgments

The CUPS Project was funded by the National Science Foundation (under grant number PHY-9014548), and it has received support from the IBM Corporation, the Apple Corporation, and George Mason University.

2

Fourier Analysis and Fourier Transforms

Brian W. James

2.1 Introduction

The use of Fourier analysis and Fourier transforms is often regarded as difficult because the explanation of the underlying process involves so much mathematics. The program, FOURIER, described in this chapter uses graphical means to illustrate the fundamental concepts of Fourier methods. Fourier analysis is illustrated by the successive summation of a series of harmonic sine and cosine waves to progressively approach the target waveforms such as a square wave (top-hat function) and a sawtooth wave.

The concept of a Fourier transform can be developed as an extension of Fourier series analysis by viewing the spectrum of amplitudes of sine and cosine harmonics for pulse waveforms using a range of narrow pulses having differing widths. The one-dimensional Fourier transforms may be investigated interactively in the program by using a slider. The similarities in spectra that are displayed show the link between Fourier series and Fourier transforms.

The program also includes a number of applications illustrating the use of two-dimensional Fourier transforms. The equivalence of Fraunhofer diffraction patterns to one- and two-dimensional Fourier transforms is shown in the text and is illustrated for a range of simple apertures in the program. The application of this equivalence is extended to image processing by the calculation of the two-dimensional Fourier transform of photographic images and the subsequent filtering of these images by setting some frequencies to zero amplitude.

The mathematical analysis, which is essential to a full understanding of the topic, is dealt with here in Appendix 2.9.1 of this chapter, but the essence of the interactive program is to show graphically how the Fourier methods provide a new insight into many phenomena and a deeper understanding of many physical processes. The figures in this chapter provide a static record of some of the results that may be obtained.

The assumption of ideal simple continuous harmonic sources for wave motion is a convenient approximation for real sources of waves, and wave effects such as interference can be described using these ideas. In reality all wave motions are limited either in time or by spatial extent so that the simple sine and cosine wave motion solutions are not sufficient. However, it is possible, using Fourier methods, to describe the more complex waveforms that occur in such phenomena as the formation of wave pulses or the dispersion of velocity. The same methods can also be applied to the effects of apertures and other phenomena that arise from the non-ideal environment in which real waves are generated and propagated.

The amplitude coefficients of sine and cosine harmonics form the Fourier series representation of an arbitrary periodic waveform. To represent a non-repetitive process in this way it is necessary to sum sine and cosine terms of all frequencies. The spectrum of these amplitudes forms the Fourier transform representation of the non-repetitive process. Fourier techniques are widely used in physics, chemistry, engineering and other physical sciences as discussed by Champeney[1] for phenomena involving time and frequency, length and wavenumber, and other reciprocal pairs of parameters. Computerized tomography, which is described extensively by Natterer,[2] and Fourier transform spectroscopy, which is discussed in Bell,[3] are excellent examples of techniques that would not be feasible without the use of Fourier transforms.

2.2 Fourier Analysis

The general form of representation of a periodic function, $f(t)$, of period T, by a Fourier series is given by Guenther[4] as

$$f(t) = \frac{a_0}{2} + \sum_{n=1}^{\infty} a_n \cos(n\omega t) + \sum_{n=1}^{\infty} b_n \sin(n\omega t), \tag{2.1}$$

where the expansion is in terms of sine and cosine functions that are harmonics of the angular frequency $\omega = 2\pi/T$. The nth harmonic of the fundamental frequency ω has amplitudes a_n for the cosine term and b_n for the sine term. The Fourier theorem states that if a function $f(t)$ is periodic, has a finite number of points of ordinary discontinuities and has a finite number of maxima and minima in the interval representing the period, then the function can be represented by a Fourier series as given in Eq. 2.1.

It is sometimes convenient to use other forms for the Fourier series and it can be shown that

$$f(t) = \sum_{n=0}^{\infty} (a_n \cos n\omega t + b_n \sin n\omega t), \tag{2.2}$$

$$f(t) = \frac{a_0}{2} + \sum_{n=1}^{\infty} c_n \cos(n\omega t + \delta_n), \tag{2.3}$$

$$f(t) = \frac{a_0}{2} + \sum_{n=1}^{\infty} d_n \sin(n\omega t + \theta_n), \tag{2.4}$$

and

$$f(t) = \sum_{n=-\infty}^{\infty} g_n e^{in\omega t}. \tag{2.5}$$

The coefficients $c_n d_n$ and g_n represent the differing amplitudes of the harmonics and δ_n and θ_n are the relevant phase shifts. It can be seen that each harmonic involves the values of two parameters. The coefficients are related by the following expressions

$$g_n = \frac{a_n + ib_n}{2} \quad (n \geq 1), \tag{2.6}$$

$$g_{-n} = \frac{a_n - ib_n}{2} \quad (n \geq 1), \tag{2.7}$$

$$(c_n)^2 = (d_n)^2 = (a_n)^2 + (b_n)^2 = 4g_n g_{-n}, \tag{2.8}$$

$$\tan \delta_n = -\frac{b_n}{a_n} \tag{2.9}$$

and

$$\tan \theta_n = \frac{a_n}{b_n}. \tag{2.10}$$

There is ambiguity about the phases δ_n and θ_n for $n \geq 1$ and about the sign for a_n, b_n, c_n, and d_n for $n \geq 1$.

2.2.1 The Coefficients for the Harmonics

To apply the Fourier theorem to a given periodic function it is necessary to determine the cosine and sine coefficients a_n and b_n. The coefficient a_0 is called the DC term because it is associated with zero frequency. Since the sine of zero is zero, there is no b_0 coefficient. It is shown in the appendix in Eq. 2.42 that

$$\frac{a_0}{2} = \frac{\omega}{\pi} \int_{-\frac{\pi}{\omega}}^{+\frac{\pi}{\omega}} f(t) dt. \tag{2.11}$$

Hence, the constant a_0 is determined by the average value of $f(t)$ over the period T.

The cosine coefficients a_m for $m \geq 1$ are obtained from Eq. 2.49 in the Appendix:

$$a_m = \frac{\omega}{\pi} \int_{-\frac{\pi}{\omega}}^{+\frac{\pi}{\omega}} f(t) \cos(m\omega t) dt. \tag{2.12}$$

The coefficients of the sine series are obtained in the appendix from Eq. 2.53:

$$b_m = \frac{\omega}{\pi} \int_{-\frac{\pi}{\omega}}^{\frac{\pi}{\omega}} f(t) \sin(m\omega t) dt. \tag{2.13}$$

Table 2.1   Fourier series expansions for some simple functions

Waveform	Coefficient value	
Square wave	$a_m = 0$	For $m = 0, 2, 4, 6\dots$
	$a_m = \dfrac{4}{\pi m}$	For $m = 1, 3, 5, 7\dots$
	$b_m = 0$	For all values of m
Triangle wave	$a_m = 0$	For $m = 0, 2, 4, 6\dots$
	$a_m = \dfrac{8}{\pi^2 m^2}$	For $m = 1, 3, 5, 7\dots$
	$b_m = 0$	For all values of m
Sawtooth wave	$a_m = 0$	For all values of m
	$b_m = \dfrac{2}{\pi m}$	For $m = 1, 2, 3, 4\dots$
Half wave rectified	$a_0 = \{1/\pi\}$	
	$a_m = 0$	For $m = 1, 3, 5, 7\dots$
	$a_m = -\dfrac{2}{\pi(m^2-1)}$	For $m = 2, 4, 6, 8\dots$
	$b_m = 0.5$	For $m = 1$
	$b_m = 0$	For $m = 2, 3, 4, 5\dots$
Full wave rectified	$a_0 = \{2/\pi\}$	
	$a_m = 0$	For $m = 1, 3, 5, 7\dots$
	$a_m = \dfrac{4}{\pi(m^2-1)}$	For $m = 2, 4, 6, 8\dots$
	$b_m = 0$	For all values of m
Ratio of pulse-width	$a_0 = -(1 - 2r)$	
to pulse-separation, r	$a_m = =\dfrac{4}{\pi m} \sin(m\pi r)$	For all values of m
	$b_m = 0$	For all values of m

2.2.2   Method of Calculation of Fourier Coefficients

In the program two methods are used for obtaining values of the coefficients. For some functions there are simple analytical solutions to the integrations and it is possible to express the coefficients in terms of a series expansion. For many functions it is more convenient to use a numerical technique to evaluate the integrations needed for each coefficient. There are some functions where algebraic integration is not possible and for these functions the use of numerical integration methods is essential. An example of this occurs in the program FOURIER when a freehand curve is Fourier analyzed following input by the cursor keys or mouse. The series expansions of the coefficients for those functions which have been integrated analytically are given in the Table. 2.1

The screen from the program in Figure 2.1 shows the Fourier analysis option, while the target square wave is being built up of the harmonics of different amplitudes. The figure shows the fundamental and the third harmonic before they are added together. The amplitude of the second harmonic is zero.

The screen in Figure 2.2 shows the build-up of the Fourier series representation of the square wave function after the first 16 terms have been summed. The seventeenth harmonic is displayed and the amplitude of the sine term is zero and the amplitude of the cosine term is 0.075. This value has been calculated in the program from the expression $a_m = \dfrac{4}{\pi m}$ with $m = 17$ given in Table 2.1 above.

It can be seen from the table of analytical solutions that cosine and sine series can be used individually to represent some functions. Consider the case of the square wave, an even function, where $f(t) = f(-t)$; then $f(t)$ can be represented

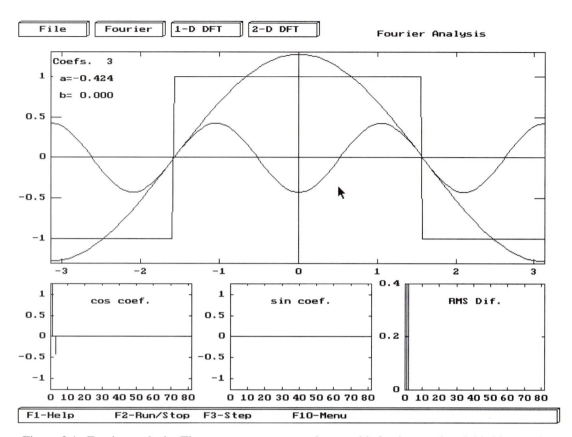

Figure 2.1: Fourier analysis: The target square wave shown with fundamental and third harmonic.

by a series of cosine waves of different amplitudes. This happens because the integral over the period from $-\pi/\omega$ to π/ω (about zero) of an even function is non-zero, but the integral of an odd function, such as the sawtooth function, over the same interval is zero. For an odd function $f(t) = -f(-t)$. Remember that the sine function is an odd function of t and that the product of an odd function with an even function is an odd function. If $f(t)$ is neither odd nor even, then both sine and cosine series are required to represent the function.

2.2.3 Root Mean Square Difference

An important aspect of Fourier analysis is the fidelity with which the summed Fourier series represents the target function. The root mean square difference of the calculated curve from the target function curve is evaluated as each harmonic is added to the series representation of the function. A histogram is plotted of the values of the root mean square difference (RMS diff.) at the lower right of the Fourier analysis display. It will be seen that the RMS difference histogram is much smaller for continuous functions than it is for functions that have discontinuities. Compare the histogram obtained with the triangle waveform to that obtained with the sawtooth waveform. It is interesting to note that the maximum difference of the Fourier series curve to that for a target top-hat function remains constant at

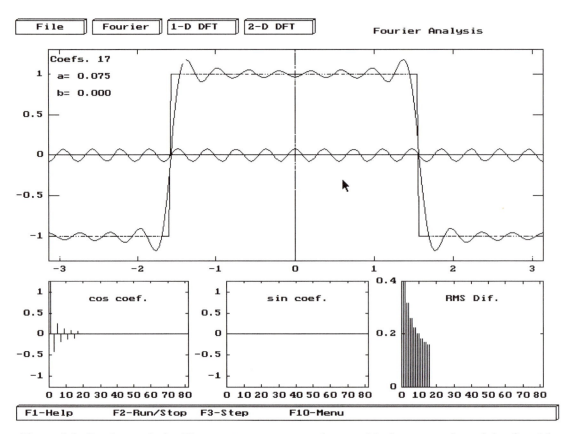

Figure 2.2: Fourier analysis: The target square wave shown with the summation of the first 16 harmonics and 17th harmonic.

about 9% of the amplitude of the square waveform. The main source of the larger RMS differences for functions with discontinuities is the difficulty of summing the continuous sine and cosine functions to produce the discontinuities. The use of sine and cosine functions leads to overshoot at each side of the discontinuity. The amount of overshoot does not tend to zero with increasing number of summed harmonics and has been shown to be about 9%.[6] This effect is called the Gibbs phenomenon.

2.3   Exponential Form of Fourier Series

It was noted earlier that other forms of representation of Fourier series were sometimes appropriate (Eqs. 2.2 to 2.5). The exponential representation is particularly useful to consider in more detail. The identity

$$\cos (n\omega t) = \frac{1}{2}(e^{i\omega t} + e^{-i\omega t}) \tag{2.14}$$

and the identity

$$\sin (n\omega t) = \frac{1}{2i}(e^{i\omega t} - e^{-i\omega t}) \tag{2.15}$$

are used to replace the cosine and sine terms in Eq. 2.1 so that

$$f(t) = \frac{a_0}{2} + \frac{1}{2}\sum_{n=1}^{\infty}(a_n - ib_n)e^{i\omega t} + \frac{1}{2}\sum_{n=1}^{\infty}(a_n + ib_n)e^{i\omega t}. \tag{2.16}$$

The coefficients of the summations in Eq. 2.16 are given by

$$\alpha_{\mp n} = a_n \pm ib_n = \frac{\omega}{\pi}\int_{-\frac{\pi}{\omega}}^{\frac{\pi}{\omega}} f(t)(\cos{(n\omega t)} \pm i\sin{(n\omega t)})dt. \tag{2.17}$$

Therefore

$$\alpha_n = \frac{\omega}{\pi}\int_{-\frac{\pi}{\omega}}^{\frac{\pi}{\omega}} f(t)e^{\pm in\omega t}dt. \tag{2.18}$$

The summations in Eq. 2.16 can be reduced to one summation from $-\infty$ to $+\infty$ by use of the representation

$$f(t) = \sum_{n=-\infty}^{\infty} \alpha_n e^{in\omega t}, \tag{2.19}$$

where

$$\alpha_n = \frac{\omega}{2\pi}\int_{-\frac{\pi}{\omega}}^{\frac{\pi}{\omega}} f(t)e^{-\{in\omega t\}}dt. \tag{2.20}$$

In the same way that the various Fourier series in Table 2.1 were classified as odd, even, or neither odd nor even according to the values of the series coefficients, then the Fourier series can be classified in terms of the a coefficient in the exponential representation as follows: for $a_n = a_{-n}$, then $f(t)$ is even; for $a_n = -a_{-n}$, then $f(t)$ is odd; for $a_n \neq \pm a_{-n}$, $f(t)$ is neither odd nor even; and for a_n complex, then $f(t)$ is complex.

2.4   The Fourier Transform

While the Fourier expansion in terms of cosine and sine series is applied to periodic functions, it is possible to treat non-periodic functions by realizing that a non-periodic function is a periodic function with infinite period. The approach to infinite period can be investigated by using the program FOURIER for a pulse signal by reducing the pulse-width to pulse-separation ratio of the pulse in the **Pulse** menu option under the main menu heading **Fourier**. The Fourier transform allows a non-periodic function to be expressed as an integral sum over a continuous range of frequencies. Thus, following Champeney,[1] for a function $f(t)$, which represents a quantity that varies with time,

$$f(t) = \int_0^{\infty} (A(\omega)\cos{(\omega t)} + B(\omega)\sin{(\omega t)})d\omega \tag{2.21}$$

or

$$f(t) = \int_{-\infty}^{+\infty} G(\nu)e^{i2\pi\nu t} d\nu. \tag{2.22}$$

The essential step in a Fourier analysis is the determination of $A(\omega)$ and $B(\omega)$ or $G(\nu)$. The functions $A(\omega)$, $B(\omega)$ and $G(\nu)$ are known as the Fourier transforms of $f(t)$. It is unfortunate that there is no universally accepted convention for the choice of which of these is the Fourier transform. This is particularly so as the factors $1/\sqrt{2\pi}$, $\frac{1}{2}\pi$ and 1 appear in the final expressions for $f(t)$. These differences arise from the use of angular frequency ω in place of frequency ν. The use of $1/\sqrt{2\pi}$ is the consequence of an attempt at symmetry.

It is a simple extension of the evaluation of the coefficients of the Fourier series in Eqs. 2.12 and 2.13 above to put

$$A(\omega) = \int_{-\infty}^{+\infty} f(t)\cos(\omega t)dt \tag{2.23}$$

and

$$B(\omega) = \int_{-\infty}^{+\infty} f(t)\sin(\omega t)dt. \tag{2.24}$$

$A(\omega)$ and $B(\omega)$ are known as the Fourier cosine and sine transforms of the function $f(t)$. In exponential representation the corresponding form for $G(\nu)$ is

$$G(\nu) = \int_{-\infty}^{+\infty} f(t)e^{-i2\pi\nu t} dt \tag{2.25}$$

and $G(\nu)$ is called the Fourier integral transform of $f(t)$.

It can be shown that $G(\nu)$ is related to $A(\omega)$ and $B(\omega)$ through the equations

$$A(\omega) = G(\nu) + G(-\nu) \tag{2.26}$$

and

$$B(\omega) = i(G(\nu) - G(-\nu)). \tag{2.27}$$

2.4.1   The Inverse Fourier Transform

The transformation from a temporal to a frequency representation in Eq. 2.25 does not result in a loss of information and it is therefore possible to perform the inverse transformation to obtain the original function. Thus Eqs. 2.22 and 2.25 define a Fourier transform pair of $f(t)$ and $G(\nu)$. The non-periodic function $f(t)$ is represented by an infinite number of sinusoidal functions with frequencies infinitely close together. $G(\nu)$ is a measure of the spectral density or the fractional contribution of frequency ν to the representation of the function. The magnitude of $G(\nu)$ is called the spectrum of the function $f(t)$.

2.4.2 Power Spectrum

In Fourier analysis the histograms of sine and cosine coefficients are plotted to show the spectral content of the cyclic function analyzed. The Fourier transform $G(\nu)$ involves real and imaginary parts and these can be plotted to show their spectral content. Another convenient pair of variables to plot is phase and amplitude. In many physical measurements the instrumentation used is not capable of resolving amplitude and phase data but can only respond to the time average of the energy that is arriving at the detector. In optics photographic film and photo-electric detectors respond to the intensity of the incident light rather than to its amplitude and phase. Thus, in order to make comparisons between theory and experiment, it is necessary to calculate the power spectrum $|G(\nu)|^2$ to obtain the variation of intensity with frequency, from the Fourier transform, which contains the amplitude and phase data in the real and imaginary parts. However, it is not possible to obtain the inverse transform as it is not possible to calculate phase and amplitude data from the power spectrum since it does not contain phase information.

The discussion above has been in terms of time and frequency. The time variable could be replaced by the space variable, x, in which case the transform variable would have reciprocal units of distance (1/distance) or spatial frequency.

2.5 Evaluation of the Fourier Integral Transform

The analytical form of some simple Fourier transforms is given in Table 2.2. Not all functions can have their transforms evaluated conveniently by analytical methods. In some situations the functional form is not known in analytical terms but only in the form of sampled data of a continuous function. For functions of this type and where digital computer programs involving numerical methods are required, it is necessary to use an approximation known as the discrete Fourier transform, or DFT, as discussed by Elliot and Rao[5] and also by Cartwright.[6]

In the DFT instead of having values of $f(t)$ at all t there are only values of $f(t)$ at a finite number of points, N,

$$0, t_1, t_2, \ldots, t_{N-1} \tag{2.28}$$

and the sample values in the sequence are

$$f(0), f(\tau), f(2\tau), \ldots, f((N-1)\tau). \tag{2.29}$$

The precision of the determination of the values of $f(t)$ will be limited either by the method of measurement, in the case of experimental data, or by the method used for the storage of the data, in the case of a computer program. This does not usually lead to significant problems in obtaining the Fourier transform, but any limitation should be considered when interpreting the Fourier transform. The number of samples obtained is usually a more significant factor in the interpretations of Fourier transforms.

Table 2.2   Fourier transforms of some simple functions.

Function	Transform
Rectangular pulse	
$f(t) = A$ for $(b - a) < t < (b + a)$	$G(\nu) = A\left(\dfrac{\sin(2\pi a \nu)}{\pi \nu}\right)e^{-i2\pi b \nu}$
$f(t) = 0$ for all other values of t	
Two rectangular pulses $b > a$	
$f(t) = A$ for $(b - a) < \mid t \mid < (b + a)$	$G(\nu) = \dfrac{2A\cos(2\pi b \nu) + \sin(2\pi a \nu)}{\pi \nu}$
$f(t) = 0$ for all other values of t	
Triangular pulse	
$f(t) = A\left(1 - \dfrac{\mid t \mid}{a}\right)$ for $-a < t < a$	$G(\nu) = aA\left(\dfrac{\sin(\pi a \nu)}{\pi a \nu}\right)^2$
$f(t) = 0$ for all other values of t	
Double-sided exponential	
$f(t) = Ae^{-a\mid t \mid}$ for $-\infty < t < \infty$	$G(\nu) = \dfrac{2A}{a}\dfrac{a^2}{a^2 + 4\pi^2 \nu^2}$
Single-sided exponential	
$f(t) = Ae^{-at}$ for $t > 0$	$G(\nu) = A\dfrac{a - i2\pi\nu}{a^2 + 4\pi^2 \nu^2}$
$f(t) = 0$ for $t < 0$	
Gaussian	
$f(t) = Ae^{-a^2 t^2}$ for $-\infty < t < \infty$	$G(\nu) = A\dfrac{\sqrt{\pi}}{a}\,exp\left(\dfrac{-\pi^2\nu^2}{a^2}\right)$
Cosine pulse	
$f(t) = \cos(bt)$ for $-a < t < a$	$G(\nu) = A\left(\dfrac{\sin(2\pi a(\nu - 1/b))}{2\pi(\nu - 1/b)} + \dfrac{\sin(2\pi a(\nu + a/b))}{2\pi(\nu + 1/b)}\right)$
$f(t) = 0$ for $\mid t \mid > a$	

In order to calculate the Fourier transform for a set of sampled points the finite form of Eq. 2.25 is used,

$$G(\nu) = \frac{1}{N}\sum_{n=0}^{N-1} f(n\tau)e^{-i2\pi\nu N\tau}, \qquad (2.30)$$

where the summation has replaced the integration. In Eq. 2.30 there are an infinite number of frequencies ν. For a fixed τ the function $e^{i2\pi\nu n\tau}$ is a periodic function of ν with a period of $2\pi/n\tau$. Thus the sum in Eq. 2.30 is periodic with period $2\pi/\tau$. It is therefore appropriate to restrict attention to the range $0 \leq 2\pi\nu \leq 2\pi/\tau$ or $0 \leq \nu \leq 1/\tau$, so that corresponding to the t domain sample there is a ν domain sample

$$0, \frac{1}{N\tau}, \frac{2}{N\tau}, \dots \frac{(N-1)}{N\tau}. \qquad (2.31)$$

Thus, the discrete Fourier transform is

$$G_{N,\tau}f\left(\frac{k}{N\tau}\right) = \frac{1}{N}\sum_{n=0}^{N-1} f(n\tau)\exp\frac{-i2\pi nk\tau}{N}. \qquad (2.32)$$

The inverse discrete Fourier transform (IDFT) of the sequence in Eq. 2.31 gives the original sequence given in Eq. 2.29 by the operation

$$f(n\tau) = \sum_{k=0}^{N-1} G_{N,\tau}f(k/N\tau)\exp\frac{i2\pi nk}{N}. \qquad (2.33)$$

The transform pair is indicated by the notation $f(n\tau)$ and $G_{N,\tau}$. The operations in Eqs. 2.30 and 2.33 are linear, and thus the DFT is a linear transformation. The function $\mid G_{N,\tau}f(n)\mid^2$ is the discrete power spectrum of the function f. The numerical procedure used to perform these transformations in the most efficient manner is known as fast Fourier transform (FFT). The one-dimensional FFT used in this program is derived from that of Press et al.[7]

The screen in Figure 2.3 shows the one-dimensional discrete Fourier transform for the **Pulse** option from the **1-D DFT** pull-down menu. The upper graph shows the original function and the lower graph shows the DFT of the pulse. The input data is generated by dividing the range $-\pi$ to $+\pi$ into 512 equally spaced intervals. The two sliders can be used to vary the width of the pulse and its position with respect to the origin. The display is updated at each movement of the sliders. The lower graph shows the DFT at 256 positive and negative harmonics. With the pulse positioned symmetrically about the origin, (the position slider at zero), it is possible to observe the similarity of the frequency distribution in this DFT to that of the Fourier analysis spectrum for a pulse. It is interesting to note that the real and imaginary parts of the DFT change as the position of the pulse is moved from the origin but that the power spectrum remains unchanged.

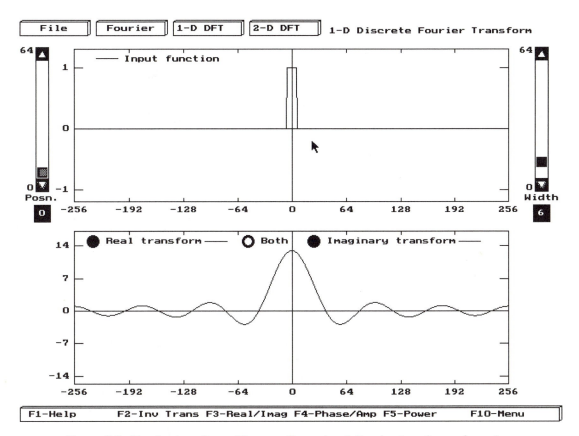

Figure 2.3: Fourier transform: The one-dimensional Fourier transform of a pulse.

2.5.1   Aliasing

It can be shown[6] that for a real-valued function the $(N-k)$th term of the DFT contains the same information as the kth term. This phenomena is called aliasing. As a consequence of aliasing it is impossible to distinguish the angular frequencies $2\pi k/N\tau$ and $2\pi(N-k)/N\tau$. This means that a frequency $2\pi(N-k)/N\tau$ higher than the sampling rate can masquerade as a component of angular frequency $2\pi k/N\tau$. If it is known what frequencies are involved in the function or the expected frequencies are too high, the data signal must be filtered before sampling with a low-pass filter of cut-off frequency $\omega_c = \pi\tau$. However, for a complex-valued function the information in the upper half of the frequency spectrum does not repeat that in the lower half.

2.5.2   Windowing

An important consequence of the cyclic nature of the DFT, defined by Eq. 2.30 is that for some functions this leads to an artificial jump at the endpoints of the data sequence.[6] The transform of these functions will contain information about the discontinuity that is an artefact of the sample window selected. It is possible to reduce the discontinuity by multiplying the sampled data by a weighting function which reduces the significance of the data in the region of the discontinuity. A triangular or cosine windowing function may conveniently be used.

2.5.3   Two-Dimensional Discrete Fourier Transforms

The concept of the Fourier transform of a function can be extended from the one-dimensional process to linear operation on a two-dimensional function that produces two-dimensional Fourier transforms. There are FFTs for discrete two-dimensional Fourier transforms given in Press et al.[7] but it is necessary to use the one-dimensional DFT method in the program because the amount of memory required to store both the real and imaginary data in a single array exceeds the size of array that can be defined in Borland Pascal. The one-dimensional transform is applied to each row of the two-dimensional array of data to produce an intermediate array. The one-dimensional Fourier transform is then used for each column of the intermediate array to produce the transform for two-dimensional array of data as explained by Press et al.[7]

The screen in Figure 2.4 shows the real part of the two-dimensional DFT of a two-dimensional pulse in which a square region around the origin is set to $+1$ while all other points are set to zero. This corresponds to a square aperture in a screen and the Fourier transform corresponds to the Fraunhofer diffraction pattern for the aperture when it is illuminated by monochromatic light. The graph at the top right is a contour plot of the diffraction surface shown in the two-dimensional projection. The orientation of the projection shown can be adjusted by using the sliders adjacent to the bottom right hand display. This display shows the current orientation as the sliders are moved. The new orientation will be used when the main display is re-calculated or re-plotted by use of the function keys **F4-Inv Trans**, **F6-Real Part**, **F7-Imaginary** or **F8-Power/gray**.

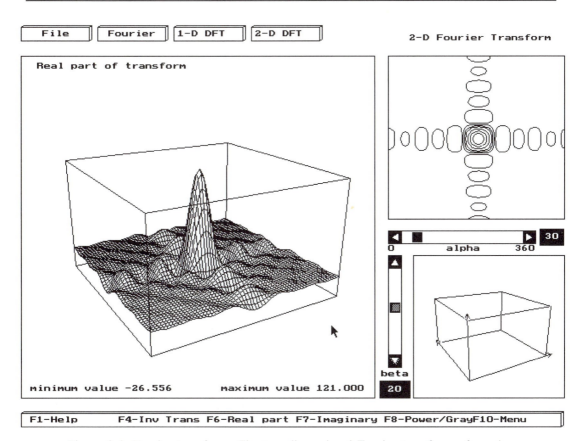

Figure 2.4: Fourier transform: The two-dimensional Fourier transform of a pulse.

2.6 Fourier Transforms and Fraunhofer Diffraction

Fraunhofer diffraction occurs when the incident waves on an aperture and the diffracted waves from the aperture are effectively plane waves. This will be the case with the source at a large distance from the aperture and the viewing point at a large distance on the other side of the aperture. In practice the plane waves requirement can be satisfied by the use of a lens on each side of the aperture so that the source and viewing plane can be at finite distances.

The general problem of diffraction by an aperture of arbitrary shape, arbitrary transmission, and arbitrary phase is considered by Fowles[8] and by Gaskill.[9] Suppose that the coordinate origins are on a common axis, as shown in Figure 2.5, that xy is the aperture plane, and that XY is the diffraction image plane in the focal plane of a lens. According to elementary geometrical optics, all rays leaving the diffraction aperture in a given direction specified by direction cosines α, β, and γ are brought to a common focus. Let this focus be located at $P(X, Y)$ where $X = L\alpha$, $Y = L\beta$ for a lens of focal length L. It is assumed here that $\tan \alpha = \alpha$, $\tan \beta = \beta$, and $\gamma \simeq 1$. The path difference δr between a ray starting from the point $Q(x, y)$ and a ray parallel to it from the origin 0 is given by $\boldsymbol{R.n}$ where $\boldsymbol{R} = \boldsymbol{i}x + \boldsymbol{j}y$ and $\boldsymbol{n}$ is

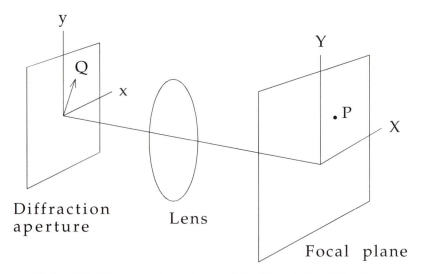

Figure 2.5: The general arrangement for Fraunhofer diffraction.

the unit vector in the direction of the ray. Clearly

$$n = i\alpha + j\beta + k\gamma, \tag{2.34}$$

then

$$\delta r = R.n = x\alpha + y\beta \; ; \tag{2.35}$$

therefore

$$\delta r = \frac{xX}{L} + \frac{yY}{L}. \tag{2.36}$$

Using the fundamental diffraction integral

$$U_p = C \int \int e^{ikr} dA, \tag{2.37}$$

where dA is an element of area about the point Q and C is the appropriate constant factor. The Fresnel-Kirchoff formula is then

$$U(X,Y) = C \int \int \exp \frac{ik(xX + yY)}{L} dxdy \tag{2.38}$$

where $k = 2\pi/\lambda$ and λ is the wavelength of light used. For a non-uniform aperture the aperture function $g(x,y)$ multiplied by $dxdy$ defines the amplitude of the diffracted wave originating from the element of area $dxdy$. Thus,

$$U(X,Y) = \int \int g(x,y)e^{i(\mu x + \nu y)} dxdy, \tag{2.39}$$

where $\mu = kX/L$ and $\nu = kY/L$. μ and ν are called spatial frequencies and have dimension of reciprocal length or wavenumber. Thus, Eq. 2.39 can be re-written as

$$U(\mu,\nu) = \int \int g(x,y)e^{i(\mu x + \nu y)} dxdy. \tag{2.40}$$

It is seen that $U(\mu,\nu)$ and $g(x,y)$ constitute a two-dimensional Fourier transform pair. The Fraunhofer diffraction pattern of an aperture is seen to be a Fourier transform representation of the aperture. Thus it is possible to investigate diffraction phenomena by obtaining the Fourier transform of the aperture. The diffraction patterns for one-dimensional apertures such as the single slit or two parallel slits may be obtained by using the one-dimensional Fourier transform for a sample of the aperture perpendicular to the axis of symmetry. Thus the double slit aperture is represented by the double delta function option under the **1-DFT** menu heading.

2.7 Running the Program FOURIER

2.7.1 Fourier Analysis

There are three main sections to the program and these are indicated by the menu headings **Fourier**, **1-D DFT** and **2-D DFT**. The **File** option included in the menu headings allows the user to load and save data in various ways, as discussed below, after selecting one of the three main options. The menu item selects the option. The function to be treated may be selected from the list in the pull-down menu. Each pull-down menu starts with some standard functions and is followed by other relevant input data choices.

- **Fourier** The hot keys **F2-Run** and **F3-Step** allow the user to build up the summed Fourier series either automatically or step-by-step. In each case the successive terms are displayed and then added in to the summed curve. The root mean square deviation from the target function of the summed terms is displayed at the bottom of the screen. Histograms of the sine and cosine coefficients are also diplayed at the bottom of the screen.

  - **Square wave, Triangle wave, Sawtooth wave, Half wave Rect,** and **Full wave Rect**: A selection of standard functions which can be expressed as a Fourier series by analytical expressions, some of which are given in Table 2.1. These analytical expressions are used in the program to calculate the Fourier sine and cosine coefficients by the use of pre-defined functions in the program.

  - **Pulse**: An input screen is provided so that the user can select the width of the pulse function for which the Fourier terms are to be calculated.

  - **Input Function, Draw Curve**, and **Load Curve Points**: The evaluation of the sine and cosine coefficients for these options for Fourier analysis involves the numerical integration of the supplied variation as given by Eqs. 2.11–2.13. The number of terms of the Fourier series that are evaluated is restricted by the number of data points that are supplied by the user in the case of the **Load Curve Points** option. As a consequence of using numerical integration, coefficients of terms which should be zero will often have very small non-zero values. This is apparent when the **Set Coefficients** option is used to set, observe, change, or modify the Fourier series that has been calculated by the program.

  - **Input Function**: This option invokes a function parser which converts the text input into a form suitable for the Pascal code. The parser uses variable x in place of the time variable t that has been used in the above description. The functions to be used should be given in a form suitable for the parser and may look a little strange compared to the normal algebraic form normally used. In particular the Greek symbol π is written as pi for the parser and the symbol $\wedge$ is used to raise a number to a power. Thus x^2 is written as $x \wedge 2$. The Heavyside function H which returns a value of 1 when the argument is greater than zero and zero otherwise is useful for generating wave pulses. For example, a square wave would be generated by the function $H(\cos(x))$ when used as the input for the parser. The function entered will be evaluated for the range $-\pi$ to $+\pi$ and should be suitably scaled to fit the amplitude range of -1 to $+1$. Complex notation for the function may be used since the parser can handle this format. The Fourier analysis will only be done for the real part of functions.

– **Draw Curve**: This option allows the user to draw almost any single valued curve by direct input into the computer by using the mouse pointer, and to obtain the Fourier series analysis of the function which is drawn. The curve must be drawn from left to right and any movement backwards will stop the drawing, which will only recommence when the pointer is to the right of any position previously drawn. This is useful if sudden steps in values are required. The range of the values should be from $-\mathbf{1}$ to $+\mathbf{1}$.

– **Set Coefficients**: Two main purposes are served by this option. It can be used to display the values of the sine and cosine coefficients that have been obtained in the analysis of any function and it can be used to investigate the effects of modifying the values of the coefficients and repeating the Fourier summation by using the hot keys in the usual way.

– **Propagate**: This option allows the user to investigate the effects of attenuation and dispersion on the propagation of the Fourier analyzed waveform. The waveform is shown after it has traveled to the right by moving the waveform to the left and calculating the Fourier sum for the new points that appear at the right of the screen. The Cauchy dispersion relation,[14] in a frequency dependent form is used, so that the phase of the summed Fourier terms depends upon the harmonic for each term. In the initial Fourier analysis the phase is independent of that of the harmonic involved as the range of phase displayed on the screen is from $-\pi$ to $+\pi$. The attenuation parameter is used to reduce by a constant fraction the value of the coefficients used in the calculation of each new point that appears on the right of the screen. This part of the program can be re-run for new values of the dispersion coefficients and the attenuation without the need to re-run the Fourier analysis section as the values of the coefficients calculated are retained in memory. This simulation would need further development to make it rigorous and it should be regarded only as an illustration of the effects of dispersion and, particularly, of the effect of attenuation on the propagation of a waveform.

– **Evolve**: This option is provided so that the user can observe the evolution of standing wave patterns similar to those that occur in musical instruments. The option is available after the initial Fourier analysis of the waveform and uses a maximum of 50 Fourier series coefficients. The maximum number of coefficients must be consistent with the 200 screen points at which the curve is evaluated. These values override those set in the **Display Setup** option while the **Evolve** option is used. The number of steps in the cycle of the waveform is adjustable from the input screen. The initial calculation is protracted because the contribution of each coefficient in each cycle has to be calculated at every screen point. The curve for each step in the cycle is displayed as it is calculated and the data is stored. Once all the curves for a cycle have been calculated, the display is generated from the stored data and a smoother animation will be observed. The **Input Function** option is particularly useful for considering the evolution of different waveforms. The patterns will be different for waves depending on whether the edges of the screen represent nodes or antinodes. There should be no constant term, $a_0 = 0$, for oscillations about an equilibrium position. The **Set Coefficients** option can be used to modify the values of the coefficients after they have been evaluated in the Fourier analysis calculations. Remember that Fourier analysis involves the assumption that the waveform analyzed is repetitive as this will affect the appearance of the waveforms.

– **Display Setup**: The **Fourier Analysis — Control of Display** input screen allows the user to modify the maximum number of harmonic terms that are evaluated automatically and which have their contributions added to the Fourier synthesized waveform when the hot key **F2-Run** is used. This is initially set to 40 and the user can enter a new number at the prompt **Maximum Number of Harmonics**. The choice of the number of points at which the Fourier series is evaluated is a compromise between the time taken to calculate each term at all points and obtaining adequate resolution. Larger values give better resolution but take longer to plot. The number of points should be roughly four times the number of harmonics and should be an even number. The user can enter a new number at the prompt **Number of Points**. Additional terms can be calculated for the Fourier series provided that there are sufficient points at which to plot each harmonic; a minimum of four points are required to represent a cycle of a sine or cosine wave, and a fifth point is the start of the next cycle. The Fourier analysis display screen always repeats the first point of a harmonic as the last point on the right of the screen. The range is $-\pi \leq t \leq +\pi$.

– **Load Coefficients**: The **File** heading of the menu includes two options that can be used after any **Fourier** option has been selected. The pull-down menu option **Load Coefficients** allows the user to read a set of Fourier series coefficients from a data file. The file SQWVCOEF.DAT is available as an example of a file of coefficients and will generate a square waveform. It starts with the number of coefficients in each series, and this is followed by that number of coefficients for cosine and sine values on separate lines of the file. The list starts with a_0 and b_0 and these are followed by further values up to the given end of the series. The summation of the synthesized series is obtained in the usual way by using the hot keys. There is no target function, so that the root mean square deviation histogram has no meaning and therefore is not plotted.

– **Load Curve**: The Fourier analysis option can be used to determine the Fourier coefficients for any single-valued data set that can be scaled in the range -1 to $+1$ for a variation of t in the range $-\pi$ to $+\pi$. The number of data points, which must be even, is given first and then the values of the equally spaced points are listed one per line in a file with the extension ".PTS." The sample file PROFFOUR.PTS contains a digitized profile of a face by Leonardo da Vinci from the book by Champeney.[10] To load a set of data points, representing a curve for which the Fourier series coefficients are required, first select an option from the standard function list in the pull-down menu under the **Fourier** option and then select **Load Curve** in the pull-down menu under **File**. A display of available files and directories will enable you to select the data file you require that has the extension ".PTS."

–   **Save Coefficients**: Once a set of Fourier coefficients have been calculated for a function it may be saved to a file by using the **Save Coefficients** option under the **File** heading on the menu. The format of the data thus saved is compatible with that used for the input to the program of a set of coefficients described above.

2.7.2   One-Dimensional Discrete Fourier Transforms

The one-dimensional discrete Fourier transform, 1-D DFT, option has a pull-down menu that contains a list of functions, some of which have analytical transforms that are given in Table 2.2.

•   **1-D DFT**: In the program the fast Fourier transform method, or FFT,[7] is used to obtain the Fourier transforms. The discrete form of the data and of the results is not shown by the use of a histogram because the DFT is only the numerical approximation used for the calculations and not an intrinsic feature of the transform, as it is in the case of Fourier analysis where harmonics of a fundamental frequency are used. The discrete nature of the transform and the resolution of the display is, however, taken into account in the calculations. The standard VGA screen has a width of 640 pixels and a height of 480 pixels. The discrete Fourier transform is most efficient for data sets of powers of two. The largest power of two consistent with the 640 pixel width is the ninth, and so 512 data points are used. A consequence of not using a histogram to represent the functions is that where the function changes by a step, for example, from zero to one, the curve drawn is approximated by a line that for half of the height of the step is drawn in the column of pixels which are associated with the zero value and only the remaining half of the line is drawn in the column of pixels corresponding to the one value of the function.

The calculation of the one-dimensional Fourier transform is relatively quick with the FFT method used and an almost instantaneous response is possible with a moderately powerful computer. It is possible, as a consequence of the rapid calculation, to use the sliders displayed at the left and right of the screen upon selection of the standard options from the sub-menu, to modify the function parameters such as width and position with respect to the origin. The hot keys **F3-Real/Imag**, **F4-Phase/Amp**, and **F5-Power** can be used at any time to select the format for the display of the transform or of the inverse transform. Graphs of the real and imaginary parts of the transform are displayed if the hot key

F3-Real/Imag is used and the radio-buttons on the screen can be used to show only the real or the imaginary part. If the hot key **F4-Phase/Amp** is used, the transform is displayed as a colored histogram showing phase (color) and amplitude. The power spectrum of the transform is shown when the hot key **F5-Power** is used. Note that the power spectrum is independent of the position of the function as set by the left-hand slider. Note also that for a function that is symmetrical about the origin the complex part is zero and that for an asymmetric function the complex part is non-zero. The functions in the pull-down menu only assign values to the real part of the input data.

The inverse transform may be obtained from the transformed data through use of the hot key **F2-Inv Trans**. The use of this key leads to the display of an input screen which allows the user to filter the transformed data by setting chosen ranges of frequency to zero, for both real and imaginary parts of the data before the inverse transform is calculated. The use of a filter is an option which must be selected on the input screen by use of a check-box. Once the input screen is accepted, the original discrete transform, after filtering, is displayed in the top half of the screen and the inverse discrete transform of the filtered transform is shown in the bottom half of the screen. The transform of the inverse transform may be obtained by further use of the hot key **F2-Transform**, and the inverse transform may also be filtered. The sliders are no longer available to vary the function once the inverse transform input screen has been accepted. If the inverse transform input screen is cancelled the sliders remain available for further adjustment.

- **Single pulse**: This option allows the user to obtain the transform of a pulse of variable width from a delta function to a maximum width of one-quarter of the total display width. The range for width and position are limited to allow any integer value to be set. The Fourier transform is displayed immediately below and the scale of the graph is automatically adjusted for the range of values. It will be seen that at maximum width the transform is tending to a delta function, as would be expected, and this is one reason to limit the width. The position can only take positive values; negative values are not provided, since the results would be a mirror image of the corresponding positive value.

- **Double pulse**: This option allows the user to find the discrete Fourier transform of two pulses of variable width, separation, and position. The width, separation, and position can be altered by the use of the three sliders. The power spectrum of the Fourier transforms produced should correspond to the diffraction pattern produced by the Young's double-slit experiment. The diffraction pattern for the ideal experiment can be obtained when the width of the slits is set to zero and discrete delta functions are produced. The effect of slits of finite width can be seen with all other positions of the width slider.

- **Grating**: This function corresponds to a regularly repeated square wave and the transform obtained should correspond to a discrete set of frequencies similar to those that would be obtained by Fourier analysis. Due to the mismatch that is a consequence of using a discrete Fourier transform, the transform obtained will be inaccurate in that it will indicate frequencies that were not present in the original when the wavelength of the square wave function is not a power of two. The power spectrum of the transform for the grating corresponds to the diffraction pattern produced by a diffraction grating illuminated by monochromatic light.

– **Cosine wave**: The two sliders allow the user to set the phase position and wavelength of the cosine wave function. This option is useful to demonstrate the effects of sampling and windowing when using the discrete Fourier transform. For instance if the wavelength is set to 11 pixels on the **Wlgth** slider then the real and imaginary spectrum of the Fourier transform will not show the correct spectrum because of the discontinuities that occur at the two ends of the sampling window. With a **Wlgth** value of 16 set on the slider, the spectrum is seen to consist of a set of discrete frequencies and the cosine curve is continuous at the ends of the sampling window.

– **Step down/up**: This function is interesting, as it produces a transform that has a constant real part and an imaginary part which changes sign at the origin when the step occurs at the origin, as would be expected for an odd function. Optically this effect can be achieved by a slit aperture that introduces a phase change half way across its width.

– **Sawtooth**: This is an odd function and thus the transform has a real part of zero and the imaginary part carries the spectral information when the sawtooth is centered on the origin.

– **Triangle**: The real part of the transform of this function, when the function is centered on the origin, is similar to the power spectrum for the pulse function. This indicates that if the photograph of a slit aperture is used as a diffracting screen for monochromatic light the diffraction pattern produced should be a double sided gray wedge.

– **Gaussian**: The algebraic Fourier transform of a Gaussian is another Gaussian. The half width of the transform is inversely related to the half width of the function.

– **Input function**: This option invokes a function parser which converts the text input into a form suitable for the Pascal code. The parser uses variable x in place of the time variable t that has been used in the description above. The functions to be used should be given in a form suitable for the parser and may look a little strange compared to the normal algebraic form normally used. In particular the Greek symbol π is written as pi for the parser and the symbol $\wedge$ is used to raise a number to a power. Thus x^2 is written as $x \wedge 2$. The Heaveyside function H which returns a value of 1 when the argument is greater than zero and zero otherwise is useful for generating wave pulses. The function entered will be evaluated for the range $-\pi$ to $+\pi$ and this range is mapped onto the 512 intervals (or pixels) of the discrete Fourier transform. The amplitude should be suitably scaled to fit the range of -1 to $+1$. Complex notation for the function may be used since the parser can handle this format. The function for the 1-D DFT will have both real and imaginary parts if a complex function is used. The imaginary part is zero for the standard functions in the menu.

– **Draw curve**: This option allows the user to draw almost any single-valued curve by direct input into the computer by using the mouse pointer and to obtain the Fourier transform of that function. The curve must be drawn from left to right, and any movement backwards will stop the drawing, which will only recommence when the pointer is to the right of any position previously drawn. This is useful if sudden steps in values are required. The range of the values should be from -1 to $+1$.

2.7.3 Two-Dimensional Discrete Fourier Transforms

The two-dimensional discrete Fourier transform, **2-D DFT**, option in the menu has a pull-down menu with a number of standard two-dimensional functions and diffraction apertures.

- **2-D DFT**: In this option, as explained in the text above, the one-dimensional FFT is used to carry out the two-dimensional transform of the chosen function. After a pull-down menu item has been selected, the user can see a projection of a three-dimensional plot of the chosen function. The real part of the selected function is displayed as surface and all the imaginary values are set to zero. The hot keys **F6-Real part**, **F7-Imaginary**, and **F8-Power/gray** are available to view different representations of the function chosen. The viewing position can be changed by using the α and β sliders to rotate the view of the small cube at the bottom right of the screen. The main three dimensional projection will subsequently be redrawn from the new view point. The hot key **F4-Transform** causes the Fourier transform to be calculated and the real part is displayed. The user can then display the Fourier transform in a variety of ways using the hot keys **F6-Real part**, **F7-Imaginary**, and **F8-Power/gray**. The inverse Fourier transform can be calculated by using the hot key **F2-Inv Trans**, which causes an input screen to be displayed that allows the user to filter the transformed data before the inverse transform is calculated and displayed.

    - **Single delta**: A function that has a value of 1 at the origin has a constant transform of one at all frequencies and is the discrete Fourier transform equivalent of a delta function. It will be seen that this applies in two-dimensions. This option is a special case of the range of rectangular input functions that are available in the next option.
    - **Rectangular hole**: This option allows the user to explore the two-dimensional equivalent of the square wave. The user can set the dimensions and position of a rectangular array of points at which the function has a value of 1. The Fourier transform of the chosen function can then be displayed, and the real, imaginary, and power spectrums can be viewed.
    - **Square frames** and **Square peg/hole**: In these options a number of simple two-dimensional shapes can be represented by functions that set certain geometric regions of points to a value of 1 or -1. These are based upon the square and the transforms show the fourfold symmetry of the original function.
    - **Circular hole**: This is a common shape for apertures in optical systems, so it is interesting to see the angular symmetry preserved even though the analysis is based on two orthogonal sets of points.

–   **Gaussian**: As in the one-dimensional case, the Fourier transform of a Gaussian surface is another Gaussian surface. Here it is interesting to note the apodization that may be achieved with a lens when a filter with a Gaussian transmission function is placed in front of the aperture.

–   **Checker board** and **Check pattern**: These are two simple rectangular patterns which fill the input function array and lead to a Fourier transform with relatively few frequencies with non-zero amplitude. This indicates the potential for identifying the presence of regular patterns from the Fourier transforms that are generated from them.

–   **Cosine wave**: This option allows the user to see the Fourier transform for a cosine wave, which is a double delta function, since the cosine wave consists of just one frequency. The imaginary part of the transform indicates the phase position of the cosine wave. The input function option allows the user to enter a wider range of cosine waves.

–   **Double delta**: This function allows the user to generate a "cosine" power spectrum when the transform is calculated. Thus it is seen that the Fraunhofer diffraction pattern produced by two pinholes in a screen is a $\cos^2$ intensity variation. It is shifted from the origin due to the form of the delta function. If the transform is filtered before the inverse transform is calculated, it will be found that the inverse transform is more complex than the original data. This indicates the risk of identifying spurious features when a Fourier transform is filtered.

–   **Input function**: This option invokes a function parser which converts the text input into a form suitable for the Pascal code. The parser uses x and y as the two independent variables to determine the value of the function. As for the one-dimensional discrete Fourier transform the functions to be used should be given in a form suitable for the parser and may look a little strange compared to the normal algebraic form normally used. In particular the Greek symbol π is written as pi for the parser and the symbol $\wedge$ is used to raise a number to a power. Thus x^2 is written as $x \wedge 2$. The Heaveyside function H, which returns a value of 1 when the argument is greater than zero and zero otherwise, is useful for generating discontinuities. The function entered will be evaluated for the range $-\pi$ to $+\pi$ for both x- and y-coordinates at the discrete values determined by the 64×64 array of points. The amplitude range is not restricted but could conveniently be scaled to fit the range of -1 to $+1$ to avoid very large or small numbers arising when the transform is calculated. Complex notation for the function may be used since the parser can handle this format. The function for the 2-D DFT will have both real and imaginary parts if a complex function is used. The imaginary part is zero for the standard functions in the menu.

–   **Picture**: This option allows the user to investigate the use of Fourier transforms in image processing. A file and directory list will be diplayed if the option has not previously been used during the current run of the program. A file NEWPICT.TXT contains a digitized picture from among those given by Gonzalez and Wintz.[11] The picture is stored as a **64 × 64** array of gray level values represented by the numbers 0 to 9 and the letters *A* to *V*. Once the file has been located and loaded it will remain

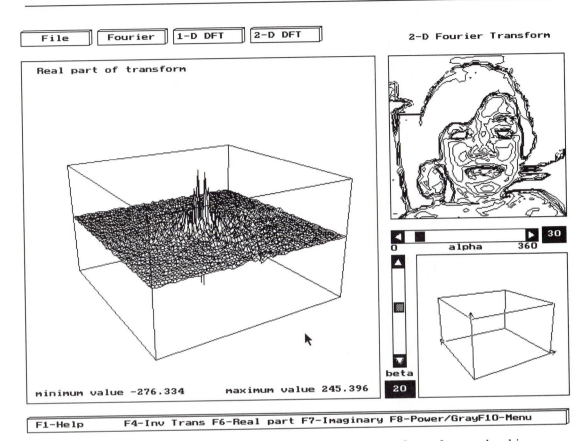

Figure 2.6: Fourier transform: The two-dimensional Fourier transform of a gray level image.

in computer memory until another picture is loaded by using the option **Load picture** under the menu item **File**. The picture will be displayed as a projection of a gray level surface and more recognizably by a contour map.

The screen in Figure 2.6 shows, in the top right corner, a contour map of the digitized gray level image in the file NEWPICT.TXT. The two-dimensional transform of the image is shown in the projection on the left of the screen. The hot key **F8-Power/gray** leads to an input screen which provides a number of other method by which the two-dimensional data may be displayed. The gray level of the real part option is most appropriate and may be selected by use of the adjacent radio-button. The 16 standard VGA colors are assigned to 16 gray levels when this display option is selected. The **64 × 64** pixel picture is displayed by a **256 × 256** array of VGA pixels. The 32 gray levels are mapped to 31 shades of gray available by using the 16 re-defined colors and by dithering to display the intermediate shades not directly available. The gray level data is converted from the single character representation to a number between −**0.5** and +**0.5** when the option is chosen so that the surface has a similar range as is displayed on the screen. The range of

−**0.5** to +**0.5** was chosen to reduce the magnitude of the zero frequency that would arise upon calculation of the Fourier transform. The data may be filtered before the inverse transform is calculated by use of the **F4-Inv trans** hot key, which leads to an input screen that allows the user to determine the "processing" that is performed on the transformed data. The use of a low-pass filter will remove high-frequency data from the image so that when it is displayed as a gray level image it will appear less sharp than before. A processed image can be saved to a file by using the **Save Picture** option under the menu heading **File**.

2.8   Exercises

2.1   Simple Functions, Even and Odd

The menu option in Fourier contains a number of simple functions. Identify those that are even and those that are odd. Confirm your identification by examining the values of the coefficients displayed in the **Set Coefficients** option of the **Fourier** menu. Why does the value of the root mean square (RMS) difference between expected and actual values tend to zero after only a few terms have been added for some waveforms?

2.2   Relationship of Fourier Coefficients to Phase of Function

Input the function $f(x) = \cos(x + k)$ with a range values of the phase constant k such as 0.1, 0.2, 0.3,... , 3.1 (π). Explain how the Fourier analysis produces sine/cosine waveforms with different phases through the coefficients of the terms in the sine and cosine series. How does your explanation help with understanding the distinction between even and odd functions and functions which are neither even nor odd? How may the same basic waveform be an even function, an odd function, or neither even nor odd?

2.3   Triangle Wave

Input the function $f(x) = (-2/pi)*ABS(x) + 1$ and compare the values of the coefficients obtained with those obtained with the menu option for a triangle wave. Why do the values of some of the coefficients obtained differ depending on the form of input of the function?

2.4   Rectified Alternating Current

Input the function $f(x) = ABS(\sin(x))$ and compare the values of the coefficients obtained with those obtained with the menu option for **Full-wave Rectification**. Why do the values of some of the coefficients obtained differ depending on the source of the function?

2.5   Sawtooth

Input the function $f(x) = (pi - x)/pi - 2 + 2*(INT(x/pi + 1))$ and compare the values of the coefficients obtained with those obtained with the menu option for a sawtooth wave. Why do the values of some of the coefficients obtained differ depending on the form of input of the function?

2.6   Pulse Width

Try a range of pulse-width to pulse-separation ratios for the pulse in the **Pulse** option in the **Fourier** menu. Note how the cosine coefficients vary in amplitude as the pulse width is reduced and tend towards a continuous spectrum of values.

2.7 A Resolution Problem

It is interesting to compare the results of carrying out the Fourier Analysis from the **Pulse** option, which are calculated analytically, with the results for the same pulse input through the **Input function** option, which uses numerical integration to obtain the coefficients. Choose the **Display setup** from the **Fourier** pull-down menu and check that the default values for the **maximum number of harmonics** is set to 40 and the **number of points** is set to 200. Select the **Pulse** option from the **Fourier** pull-down menu and set the pulse width to 0.1. Observe the graphs of the Fourier coefficients that are generated after **F2-Run** is pressed. Select the **Input function** from the **Fourier** pull down menu and enter the function $f(x) = H(\cos(x) - \cos(c * pi)) * 2 - 1$ with $c = 0.1$. Observe the graphs of the Fourier coefficients that are generated after **F2-Run** is pressed. This should generate the same set of coefficient values that were obtained with the width in the **Pulse** option set to 0.1. Compare the values of the two sets of coefficients. Remember that the coefficients can be written to a file or that the coefficient values are displayed when the hot key **F3-Step** is used to sum and display the Fourier series.

Now use the **Pulse** option with the width set to 0.01 and again observe the graphs of coefficients. Use the **Input function** option again with the same function for a value of $c = 0.01$ and observe the coefficients. Repeat the **Input function** option calculations after using the **Display setup** to change the **number of points** to 400. Comment on the values of the newly calculated coefficients compared with the previous values. Why do the values depend on the number of data points used?

Repeat the Fourier analysis of the function with $c = 0.01$ after using the **Display setup** to change the **number of points** to 100, 200, 300, 400 and 316. Comment on the values obtained for the coefficients. Try the Fourier analysis with $c = 0.02$. What would you expect the coefficient values to tend towards as c tends to zero?

2.8 Square Wave

The Heaveyside function returns a value of 1 when $f(x) > 0$ and a value of zero when $f(x) < 0$. Input the function $f(x) = 2 * H(\sin(x)) - 1$ and compare it with the **Square wave** option in the pull-down menu by inspecting the coefficients for the harmonics of the Fourier series obtained.

2.9 Wave Packet

Use the Heaveyside function to generate a pulse of waves by using the function $f(x) = 2 * H(\cos(x)) * \cos(9 * x)$. Compare the Fourier coefficients obtained with the product of the coefficients for a square wave with those for the $\cos(9 * x)$. Use the propagate option to view the effects of dispersion when the waveform is propagated for various values of the Cauchy constants.

2.10 Finger Outline

With your fingers touching each other, draw an outline of your hand onto graph paper and prepare a data file which gives the outline in digital form.

You will need about 100 equally spaced points. The data should start with the number of points, which should be followed by the y-coordinates for all the points. You may need to scale the data so that values of y lie in the range -1 to $+1$. The file PROFFOUR.PTS is an example of such a data file. You can use this program to obtain the Fourier series coefficients for any suitably prepared data.

2.11   Fourier Analysis of a Complex Function

The parser used with the **Function input** option handles complex functions so that it is possible to obtain the Fourier series coefficients for the real part of a complex function. Enter the function $f(x) = 0.5 *$ $(\exp i * 5 * x) + \exp(-i * 5 * x)$ and obtain the Fourier coefficients.

2.12   Windowing of 1-D DFT

Input the function $f(x) = \cos(7.5 * x)$ and then the function $f(x) = \cos(7 * x)$ into the **Input function** option of the pull-down menu under **1-D DFT**. Comment upon the one-dimensional discrete Fourier transform spectra that are obtained and explain why these are different from the analytical Fourier transforms for these functions. Use the **Input function** option to investigate the effect of multiplying the function $f(x) = \cos(9.5 * x)$ by the window function $\cos(x/2)$. Try the effect of using other window functions, such as a triangular function. Note particularly what happens at the edges of the sample window before and after using the cos and triangular functions.

2.13   Fourier Transform of a Slit Diffraction Image

Use the **Input function** option of the pull-down menu under **1-D DFT** to view the Fourier transforms of $\sin(16 * x)/(16 * x)$ and $(\sin(16 * x)/(16 * x))^2$. By considering these transforms explain why the amplitude of the diffraction pattern produced by a photograph of the diffraction pattern of a narrow slit has a triangular variation in ampliude and not the pulse variation of the original slit source. Remember that the power spectrum corresponds to the intensity of a diffraction pattern.

2.14   Aliasing Effects

Use the **Input function** option of the pull-down menu under **1-D DFT** to investigate aliasing effects. Input the function of the form $f(x) = \cos(n * x)$, where n is an integer greater than 256, such as 264, which will alias as 248 in the transform.

2.15   Recovering Data From Amplitude Noise With a 1-D DFT

Use the random number function, available with the parser, to simulate noise with a mean value of zero and add it to a sine or cosine function of low amplitude. On input of the function $f(x) = (rand - 0.5) + 0.1 * \cos(28 * x)$ it will be found that there is a peak in the Fourier transform at 28. This will be more clearly seen in the power spectrum.

2.16   Recovering Data From Phase Noise With a 1-D DFT

The random number function, available with the parser, can be used to simulate phase noise about a mean value. On input of the function $f(x) = \cos(32 * x + rand * pi * k)$ with $k \leq 1.5$ it will be found that there is a peak in the Fourier transform at 32. This will be more clearly seen in the power spectrum. The color of the peaks at 32 in the phase amplitude plot will indicate the mean phase shift. The mean phase shift can be altered by the addition of a constant phase shift.

2.17 Fourier Transform of a Complex Function

The parser used with the **Function input** option handles complex functions so that it is possible to obtain the Fourier transform of a complex function. Enter the function $f(x) = \cos(5 * x) - i \sin(5 * x)$. Try the effect of changing the minus sign to a plus sign. To display the imaginary part of the input function use **F2-Inv Trans** to obtain the inverse transform of the displayed Fourier transform without using the option to filter the transform.

2.18 A Smaller Array for the 2-D DFT

The two-dimensional fast Fourier transform is set to work with a **64 × 64** array of data by the constants maxRow and maxCol at the beginning of the program. Try changing the value of these two constants to 32. You should find that all the **2-D DFT** options except **Picture** work satisfactorily. It will be found that the time spent calculating for a transform is much shorter by a factor of nearly 16, as the number of calculations is proportional to the square of the number of data points.

2.19 Computer-Generated Hologram

The two-dimensional Fourier transform option provides the user with the opportunity to generate a simple hologram following the method described by Lee.[12] The first step is to generate a data file for the hologram to be generated. The simplest method would be to import the necessary data in the same way as the gray-level pictures are loaded. A simple editor would be adequate to generate the 64 × 64 array of characters needed. The coding regards 0 as black and *V* as white with the symbols 0 to 9 and *A* to *V* representing the intermediate gray levels. A modification to the program to give output of the real and imaginary transform data to a file would need to be implemented. It would also be necessary to convert this data to a hologram, which could be done by using PostScript, as described by Macgregor.[13] The hologram can be then be produced photographically. The photographic image can be illuminated by a laser to produce the final hologram image.

References

1. Champeney, D.C. *Fourier Transforms and Their Physical Applications.* London: Academic Press, 1973.

2. Natterer, F. *The Mathematics of Computerized Tomography.* Chichester: John Wiley and Sons, 1986.

3. Bell, R.J. *Introductory Fourier Transform Spectroscopy.* New York: Academic Press, 1972.

4. Guenther, R. *Modern Optics.* New York: John Wiley and Sons, 1990.

5. Elliott, D.F., Rao, K.R. *Fast Transforms Algorithms, Analyses, Applications.* New York: Academic Press, 1982.

6. Cartwright, M. *Fourier Methods for Mathematicians, Scientists and Engineers.* London: Ellis Horwood, 1990.

7. Press, W.H., Flannery, B.P., Teukolsky, S.A., Vetterling, W.T. *Numerical Recipes in Pascal.* Cambridge: Cambridge University Press, 1989.

8. Fowles, G.R. *Introduction to Modern Optics.* New York: Holt, Rinehart and Winston Inc. 1975.

9. Gaskill, J.D. *Linear Systems, Fourier Transforms and Optics.* New York: John Wiley and Sons, 1978.

10. Champeney, D.C. *Fourier Transforms in Physics.* Bristol: Adam Hilger Ltd., 1985.

11. Gonzalez, R.C., Wintz, P. *Digital Image Processing.* London: Addison-Wesley Pub. Co., 1977.

12. Lee, W.H. *Sampled Fourier Transform Hologram Generated by Computer.* Applied Optics **9**:3 639–643, 1970.

13. Macgregor, A.E. *Computer generated holograms from dot matrix and laser printers.* American Journal of Physics. **60**:9, 839–846, 1992.

14. Pedrotti, F.L. and Pedrotti, L. S., *Introduction to Optics.* Englewood Cliffs, NJ: Prentice Hall Inc., 1993.

2.9   APPENDICES

2.9.1   The Coefficients of the Cosine Terms

To apply the Fourier theorem to a given periodic function it is necessary to determine the coefficients a_n and b_n. The coefficient a_0 is called the DC term because it is associated with zero frequency. Since the sine of zero is zero, there is no b_0 coefficient. To determine a_0 both sides of Eq. 2.1 are multiplied by dt and the function is integrated over one period. It is convenient if this period is located symmetrically about the origin so that the limits of integration are $-\pi/\omega$ to π/ω. Thus,

$$\int_{-\frac{\pi}{\omega}}^{+\frac{\pi}{\omega}} f(t)dt = \int_{-\frac{\pi}{\omega}}^{+\frac{\pi}{\omega}} \frac{a_0}{2}dt + \sum_{n=1}^{\infty} \int_{-\frac{\pi}{\omega}}^{+\frac{\pi}{\omega}} a_n \cos(n\omega t)dt$$

$$+ \sum_{n=1}^{\infty} \int_{-\frac{\pi}{\omega}}^{+\frac{\pi}{\omega}} b_n \sin(n\omega t)dt. \qquad (2.41)$$

The integral of a sine or a cosine function over one period or an integral multiple of one period is zero; thus

$$\frac{a_0}{2} = \frac{\omega}{\pi} \int_{-\frac{\pi}{\omega}}^{+\frac{\pi}{\omega}} f(t)dt. \tag{2.42}$$

Hence, the constant a_0 is determined by the average value of $f(t)$ over the period T.

The coefficients a_m for $m \geq 1$ are obtained by multiplying both sides of Eq. 2.1 by $\cos(m\omega t)$ and then integrating from $-\frac{\pi}{\omega}$ to $+\frac{\pi}{\omega}$ so that for the mth term

$$\int_{-\frac{\pi}{\omega}}^{+\frac{\pi}{\omega}} f(t) \cos(m\omega t)dt$$

$$= \int_{-\frac{\pi}{\omega}}^{+\frac{\pi}{\omega}} \frac{a_0}{2} \cos(m\omega t)dt + \sum_{n=1}^{\infty} \int_{-\frac{\pi}{\omega}}^{+\frac{\pi}{\omega}} a_n \cos(n\omega t) \cos(m\omega t)dt$$

$$+ \sum_{n=1}^{\infty} \int_{-\frac{\pi}{\omega}}^{+\frac{\pi}{\omega}} b_n \sin(n\omega t) \cos(n\omega t)dt. \tag{2.43}$$

The trigonometric identities

$$\sin(n\omega t) \cos(m\omega t) = \frac{1}{2}[\sin((n + m)\omega t) + \sin((n - m)\omega t)] \tag{2.44}$$

and

$$\cos(n\omega t) \cos(m\omega t) = \frac{1}{2}[\cos((n + m)\omega t) + \cos((n - m)\omega t)] \tag{2.45}$$

are used to evaluate the two summations. The first summation involves terms of the form

$$\int_{-\frac{\pi}{\omega}}^{+\frac{\pi}{\omega}} a_n \cos(n\omega t) \cos(m\omega t)dt = \frac{1}{2} \int_{-\frac{\pi}{\omega}}^{+\frac{\pi}{\omega}} a_n \cos((n + m)\omega t)dt$$

$$+ \frac{1}{2} \int_{-\frac{\pi}{\omega}}^{+\frac{\pi}{\omega}} a_n \cos((n - m)\omega t)dt. \tag{2.46}$$

Both the integrals are zero for $n \neq m$. In the case of $n = m$ the first integral is zero, but the second integral is $(\pi/\omega)a_m$. The second summation involves terms of the form

$$\int_{-\frac{\pi}{\omega}}^{+\frac{\pi}{\omega}} b_n \sin(n\omega t) \cos(m\omega t)dt = \frac{1}{2} \int_{-\frac{\pi}{\omega}}^{+\frac{\pi}{\omega}} b_n \sin((n + m)\omega t)dt$$

$$+ \frac{1}{2} \int_{-\frac{\pi}{\omega}}^{+\frac{\pi}{\omega}} b_n \sin((n - m)\omega t)dt. \tag{2.47}$$

These terms are zero for all values of n. Therefore,

$$\int_{-\frac{\pi}{\omega}}^{+\frac{\pi}{\omega}} f(t) \cos(m\omega t) = \frac{1}{2} \int_{-\frac{\pi}{\omega}}^{+\frac{\pi}{\omega}} a_m dt = \frac{\pi a_m}{\omega}. \tag{2.48}$$

Thus, the coefficients of the cosine series are obtained from the integral

$$a_m = \frac{\omega}{\pi} \int_{-\frac{\pi}{\omega}}^{+\frac{\pi}{\omega}} f(t) \cos(m\omega t) dt. \tag{2.49}$$

2.9.2   The Coefficients of the Sine Terms

The coefficients for the sine series start with $b_0 = 0$, as the sine term makes no contribution to the average value. The coefficient b_m for $m \geq 1$ may be obtained in a similar manner to that for the coefficient a_m. Thus, multiplying Eq. 2.1 by $\sin(m\omega t)$, it is found that

$$\int_{-\frac{\pi}{\omega}}^{\frac{\pi}{\omega}} f(t) \sin(m\omega t) dt = \int_{-\frac{\pi}{\omega}}^{\frac{\pi}{\omega}} \frac{a_0}{2} \sin(m\omega t) dt$$

$$+ \sum_{n=1}^{\infty} \int_{-\frac{\pi}{\omega}}^{\frac{\pi}{\omega}} a_n \cos(n\omega t) \sin(m\omega t) dt$$

$$+ \sum_{n=1}^{\infty} \int_{-\frac{\pi}{\omega}}^{\frac{\pi}{\omega}} b_n \sin(n\omega t) \sin(m\omega t) dt. \tag{2.50}$$

The trignometric identity

$$\sin(n\omega t) \sin(m\omega t) = \frac{1}{2}(\cos((n - m)\omega t) - \cos((n + m)\omega t)) \tag{2.51}$$

is used to obtain a simple expression for b_m. In the case of the b coefficients the second term of the summation involves terms of the form

$$\int_{-\frac{\pi}{\omega}}^{+\frac{\pi}{\omega}} b_n \sin(n\omega t) \sin(m\omega t) dt = \frac{1}{2} \int_{-\frac{\pi}{\omega}}^{\frac{\pi}{\omega}} b_n \cos((n - m)wt) dt$$

$$- \frac{1}{2} \int_{-\frac{\pi}{\omega}}^{\frac{\pi}{\omega}} a_n \cos((n + m)wt) dt. \tag{2.52}$$

The first integral on the right of this equation is non-zero when $n = m$ and zero in all other cases. The second integral on the right of the equation is zero for all values of n. Thus, the coefficients of the sine series are obtained from the integral

$$b_m = \frac{\omega}{\pi} \int_{-\frac{\pi}{\omega}}^{\frac{\pi}{\omega}} f(t) \sin(m\omega t) dt. \tag{2.53}$$

3

The Wave Equation and Other Partial Differential Equations

Wolfgang Christian, Susan Fischer, Andrew Antonelli

3.1 Introduction

Partial differential equations (PDEs) are central to our understanding of physics. Wave equations model the propagation of sound, light, and water waves; the Schrödinger equation predicts the properties of atomic and molecular systems; the diffusion equation describes the flow of particles or energy. Numerical solution of a wide variety of PDEs makes it possible to demonstrate the similarities and differences between the various equations, thereby forming a bridge between the classical mechanics of the 19th century and the quantum mechanics of the 20th. It was, however, necessary to limit certain features of the program to the classical wave equation in order to limit the scope and size of the code. For instance, quantum mechanical operators are not discussed since they are treated extensively elsewhere in the CUPS series.

The reader may wish to consult a book on mathematical methods for scientists, as well as standard texts on optics, quantum mechanics, and thermal physics while working through the computer exercises accompanying program WAVE. Texts which provide extensive treatments of oscillatory and wave phenomena are also available.[3,7,10]

3.2 Classification of Partial Differential Equations

Second-order PDEs of two variables are of the form

$$a\frac{\partial^2 f(x,y)}{\partial x^2} + b\frac{\partial^2 f(x,y)}{\partial x\partial y} + c\frac{\partial^2 f(x,y)}{\partial y^2} + d\frac{\partial f(x,y)}{\partial x} + e\frac{\partial f(x,y)}{\partial y} = F(x,y) \qquad (3.1)$$

and can be classified on the basis of the discriminant, $b^2 - 4ac$, into three types based on the conic sections of the same name:

$$b^2 - 4ac > 0 \qquad \text{hyperbolic}$$

$$b^2 - 4ac = 0 \qquad \text{parabolic}$$

$$b^2 - 4ac < 0 \qquad \text{elliptic} \tag{3.2}$$

Examples of hyperbolic, parabolic, and elliptic equations are the classical wave equation, the diffusion equation, and Laplace's equation, representing the application of PDEs to sound and light, heat, and electrostatic phenomena, respectively. The type of solution for each of these equation types is quite different. Elliptic equations produce stationary and energy-minimizing solutions, parabolic equations produce a smooth-spreading flow of an initial disturbance, and hyperbolic equations produce a propagating disturbance. For example, the first PDE encountered by a student may be the one-dimensional wave equation,

$$\frac{\partial^2 u(x,t)}{\partial t^2} = v^2 \frac{\partial^2 u(x,t)}{\partial x^2}, \tag{3.3}$$

since it predicts the time evolution of many types of transverse and longitudinal waves.[2,3,4] Replacing y in Eq. 3.1 by t, we see that the classical wave equation is an example of a hyperbolic PDE. Since a suitable change of variables will reduce any hyperbolic equation to this canonical form (to first order) we are really studying a very general problem. Table 3.1 lists the PDEs that can be solved by the accompanying program, WAVE. As is typical in computational work, units will be chosen to minimize roundoff errors. Notice, for instance, that mass and Planck's constant are unity in the Schrödinger equation.

Another useful classification of partial differential equations is based on a property called *linearity*. Assume two different solutions of a differential equation, u_α and u_β, have been found. Then the equation is said to be linear if their sum

$$u = u_\alpha + u_\beta \tag{3.4}$$

is also a solution of the equation. Linearity is a very desirable property in many systems, since it allows us to express the time evolution of an arbitrary initial

Table 3.1   Partial differential equations solved by WAVE

$\frac{\partial^2 u(x,t)}{\partial t^2} - c^2 \frac{\partial^2 u(x,t)}{\partial x^2} = 0$	Classical wave
$\frac{\partial^2 u(x,t)}{\partial t^2} - \frac{\partial^2 u(x,t)}{\partial x^2} = u(x,t)$	Klein-Gordon
$\frac{\partial^2 u(x,t)}{\partial t^2} - \frac{\partial^2 u(x,t)}{\partial x^2} = \sin u(x,t)$	Sine-Gordon
$\frac{\partial^2 u(x,t)}{\partial t^2} - \frac{\partial^2 u(x,t)}{\partial x^2} = \sin u(x,t) + 1/2 \sin u(x,t)/2$	Double sine-Gordon
$\frac{\partial^2 u(x,t)}{\partial t^2} - \frac{\partial^2 u(x,t)}{\partial x^2} = -u(x,t) + u(x,t)^3$	Phi-four
$\frac{\partial u(x,t)}{\partial t} - \kappa \frac{\partial^2 u(x,t)}{\partial x^2} = 0$	Diffusion
$i \frac{\partial u(x,t)}{\partial t} + \frac{\partial^2 u(x,t)}{\partial x^2} = V(x)u(x,t)$	Schrödinger

state, $u(x, 0)$, in terms of the time evolution of some complete set of known and presumably simple functions. (See chapter 2 for a discussion of Fourier methods.) For example, if the sound of a violin and a piano are processed simultaneously by a stereo system, the listener should be able to hear the resultant music as if each instrument were played separately and the sound intensity added together. Any nonlinearity in the stereo system is viewed as an undesirable distortion. Other phenomena, such as laser-pulse compression or frequency doubling, depend on system nonlinearity. Both linear and nonlinear systems will be studied in the exercises.

3.3 Physics of Partial Differential Equations

3.3.1 Classical Wave Equation

Although the one-dimensional classical wave equation arises in many contexts, it is the interaction between time-varying electric and magnetic fields predicted by Faraday's and Ampere's laws that provides the most interesting phenomena. Consider the y-component of a time-varying electric field in vacuum, $E_y(x, t)$. The line integral along a path in the x-y plane

$$(x, y, z) \rightarrow (x + \Delta x, y, z) \rightarrow (x + \Delta x, y + \Delta y, z)$$
$$\rightarrow (x, y + \Delta y, z) \rightarrow (x, y, z) \tag{3.5}$$

can be approximated by

$$emf = \oint E_y \, dl = \Delta E_y \Delta y, \tag{3.6}$$

where $\Delta E_y = E_y(x + \Delta x, t) - E_y(x, t)$. Faraday's law tells us that the induced *emf* is equal to the rate of decrease of the magnetic flux, $\Phi = B_z \Delta x \Delta y$, through the closed path:

$$\Delta E_y \Delta y = -\frac{\partial B_z}{\partial t} \Delta x \Delta y. \tag{3.7}$$

In the limit as $\Delta x \rightarrow 0$ we obtain

$$\frac{\partial E_y}{\partial x} = -\frac{\partial B_z}{\partial t}. \tag{3.8}$$

Similarly, the line integral of the magnetic field in the z direction, $B_z(x, t)$, taken around a loop in the $x-z$ plane (the minus sign is chosen so that the right-hand rule gives positive circulation about the y-axis),

$$\oint B_z \, dl = -\Delta B_z \Delta z, \tag{3.9}$$

is proportional to the displacement current,

$$I_D = \epsilon_0 \frac{\partial E_y}{\partial t} \Delta x \Delta z, \tag{3.10}$$

by Ampere's law and yields the following equation:

$$-\Delta B_z \Delta z = \mu_0 \epsilon_0 \frac{\partial E_y}{\partial t} \Delta x \Delta z. \tag{3.11}$$

In the limit as $\Delta x \rightarrow 0$ we obtain

$$\frac{\partial B_z}{\partial x} = -\mu_0 \epsilon_0 \frac{\partial E_y}{\partial t}. \tag{3.12}$$

Although good numerical techniques are available to solve Eqs. 3.8 and 3.12 as a coupled system of first-order PDEs,[13] we will convert these equations into a single second-order equation. Taking the derivative of Eq. 3.8 with respect to x and Eq. 3.12 with respect to t and equating the mixed partial derivatives leads to the desired equation

$$v^2 \frac{\partial^2 E_y(x,t)}{\partial x^2} = \frac{\partial^2 E_y(x,t)}{\partial t^2}, \tag{3.13}$$

where $v = \sqrt{1/\mu_0 \epsilon_0}$.

An interesting feature of Eq. 3.13 is that there is a second identical equation which governs the propagation of the magnetic field. Although the computer display shows only the transverse electric field, the magnetic field is just as important and, in fact, carries an equal amount of energy. Computer exercises will ask you to examine the relationships between these fields and the energy density.[2]

3.3.2   Klein-Gordon and Schrödinger Equations

The idea that waves are associated with differential equations was well-established when Debye suggested to Schrödinger that he try to find the appropriate equation for deBroglie waves. Although it is impossible to derive Schrödinger's equation, it is possible to make plausible assumptions and then compare the results with experiment. If we assume that a free particle propagating in the x-direction can be approximated by a plane wave and adopt deBroglie's relations for momentum and energy, $p = \hbar k$ and $E = \hbar \omega$, we obtain a wave function of the form

$$\phi(x,t) = Ae^{i(kx-\omega t)} = Ae^{i(px-Et)/\hbar}. \tag{3.14}$$

Substituting Eq. 3.14 in place of the electric field into Eq. 3.13 results in the energy-momentum relationship for particles of zero rest mass, i.e., photons:

$$E^2 = p^2 c^2. \tag{3.15}$$

In order to describe particles of finite rest mass, it is necessary to modify Eq. 3.13 to be relativisticly invariant by using Einstein's mass-energy relationship

$$E^2 = p^2 c^2 + m^2 c^4. \tag{3.16}$$

These considerations led Schrödinger to the following equation in 1925:

$$\frac{\partial^2 \phi(x,t)}{\partial x^2} - \frac{1}{c^2} \frac{\partial^2 \phi(x,t)}{\partial t^2} - \frac{m^2 c^2}{\hbar^2} \phi(x,t) = 0. \tag{3.17}$$

However, this equation was first published by Klein and Gordon and is now known as the Klein-Gordon (KG) equation. It is still used in some textbooks as an introduction to relativistic wave equations.[14] What we now call *the* Schrödinger equation,

$$i\hbar \frac{\partial \Psi(x,t)}{\partial t} = \frac{-\hbar^2}{2m} \frac{\partial^2 \Psi(x,t)}{\partial x^2} + V(x)\Psi(x,t), \tag{3.18}$$

is actually the nonrelativistic limit of Eq. 3.17.[9]

3.3.3 Nonlinear Klein-Gordon Equations

Although nonlinear wave equations were known in the 19th century, only a few solutions were known and their properties were difficult to study. The most famous of these equations models a water wave propagating in a canal. It was published in 1895 by Korteweg and de Vries and is now known as the KdeV equation. Korteweg and de Vries showed that this equation has pulse-shaped solutions which propagate without changing shape. Such solutions were called solitary waves; they were considered mathematical curiosities until fairly recently. Before the advent of the computer, it was assumed that nonlinearities would cause solutions of nonlinear PDEs to interact in such a way that their identity would be lost if they overlapped. After all, $y_1 + y_2$ would probably not be a solution even though y_1 and y_2 were. However, computer experiments showed that solutions existed which would emerge from a collision with their shape and velocity unchanged. Such special solitary waves were named solitons in recognition of these particle-like properties. Many other nonlinear wave equations that exhibit solitary wave and soliton-like behavior have since been discovered.

We model the three nonlinear Klein-Gordon (NKG) equations listed in Table 3.1. Using appropriate units, these equations are of the form

$$\frac{\partial^2 \phi}{\partial t^2} - \frac{\partial^2 \phi}{\partial x^2} = F(\phi). \tag{3.19}$$

Since Eq. 3.19 is an extension of the linear KG equation, it is not surprising that it can be solved by the same numerical technique.

The linear Klein-Gordon equation is the equation for a relativistic quantum-mechanical scalar (spin-zero) particle of mass m. We have already shown that if $\phi(x,t) = Ae^{i(px - Et)}$, a plane wave, where $p = k = momentum$ and $E = \omega = energy$, then you get $p^2 + m^2 = E^2$, which is the correct relativistic relationship between rest energy, momentum, and energy in units where $\hbar = 1$ and $c = 1$. The linear KG equation can also be obtained by using the Euler-Lagrange equations from the Lagrangian density

$$L = -(1/2)\left(\frac{\partial^2 \phi}{\partial x^2} - \frac{\partial^2 \phi}{\partial t^2}\right) - (1/2)\phi^2. \qquad (3.20)$$

The function $\phi(x, t)$ now represents a quantized field describing the particle. Again, this is a *free* particle without interactions. Now consider how these particles might interact among themselves. The simplest term that can be added to the Lagrangian is a quadratic term of the form

$$\Delta L = -(1/4)b\,\phi^4, \qquad (3.21)$$

where b is a constant. The resulting wave equation is called the phi-four equation.

What does this mean? Classically, this is the same as adding a term $+(1/4)b\,\phi^4$ to the potential energy, which in the original formalism was just $+(1/2)\phi^2$ where the minimum energy state is at $\phi(x, t) = 0$. This is still true when we add the quadratic term. (A cubic term would have put the minimum energy state at $x \to \pm\infty$.) From the quantum field theoretic perspective, this term in the Lagrangian means that processes occur in which two scalar particles can scatter off each other at a point. The constant b is then a measure of the strength of the interaction between particles (i.e., a coupling constant) and has been set equal to unity in our program.

The phi-four theory as described above is a "toy" model, in that there don't appear to be real, physical particles that interact in this simple way. But it is a good way to begin understanding the overall character of field theories of interacting particles.

The sine-Gordon equation is another nonlinear extension of the KG equation. It also originated in particle field theory, although it also models a Josephson junction and is the continuum limit for a chain of pendula coupled by springs. It is the only common nonlinear NKG equation that is completely integrable; i.e., it has exact analytical solutions. However, the solutions must be written in terms of Jacobi elliptic functions. The double sine-Gordon equation is similar and has applications in nonlinear optics. All of the above NKG equations will have what are called *kink* solutions that take the system from one asymptotically stable state to another.

3.3.4   Diffusion Equation

Diffusion is an everyday experience; a hot cup of tea distributes its thermal energy or an uncorked perfume bottle distributes its scent throughout a room. Since nonuniform distributions tend to distribute themselves in such a way as to produce uniformity, it is reasonable to assume that the flux density, J, is proportional to the concentration gradient. For a one-dimensional system we can write

$$J(x, t) = -K\frac{\partial n(x, t)}{\partial x}, \qquad (3.22)$$

where n is the density of the physical quantity of interest. The proportionality constant, K, is given various names depending on the quantity that n represents.

It is called the *thermal conductivity* if n is heat and the *coefficient of diffusion* if n is molecular concentration. For the purposes of this discussion, we will consider $n(x, t)$ to be a mass density along the x-axis and we will refer to the integral of this function, $\int_a^b n(x, t)\, dx$, as the total mass.

The number of molecules between x and $x + \Delta x$ can be approximated by $n(x, t)\Delta x$. If we assume that the total number of molecules is conserved (i.e., no source terms), then the number of molecules within the interval Δx will change due to the fluxes at the two ends, x and $x + \Delta x$:

$$\frac{\partial}{\partial t} n(x, t)\Delta x = J(x, t) - J(x + \Delta x, t). \tag{3.23}$$

The above equation is the differential form of the conservation of mass. A Taylor expansion about x of the last term allows the equation to be rewritten in the form

$$\frac{\partial n(x, t)}{\partial t} = -\frac{\partial J(x, t)}{\partial x}, \tag{3.24}$$

which can be combined with Eq. 3.22 to produce the diffusion equation

$$\frac{\partial n(x, t)}{\partial t} = K \frac{\partial^2 n(x, t)}{\partial x^2}. \tag{3.25}$$

3.4 Numerical Methods

3.4.1 Finite Difference Approximation

The numerical solution of PDEs is often neglected in the undergraduate physics curriculum even though it can lead to useful insights into the underlying physics.[15,16] Their solution is hardly more difficult than the solution of ordinary differential equations, once the notation required to work with more than one variable is mastered. Since computers have a finite amount of memory, we must limit our knowledge of the solution, $u(x, t)$, to a finite number of points in space and time. This process of discretizing space and time is accomplished by letting the continuous variables x and t take on the values given by

$$x_j = x_0 + j\Delta x \qquad j = (0, 1, 2 \ldots N)$$
$$t_n = t_0 + n\Delta t \qquad n = (0, 1, 2 \ldots M)$$
$$u_j^n = u(x_j, t_n). \tag{3.26}$$

The wave amplitude $u(x, t)$ is now described by an N by M matrix, where subscripts and superscripts refer to spatial and temporal variables, respectively:

$$U = \begin{array}{c} t \uparrow \end{array} \begin{pmatrix} u_0^M & u_1^M & \cdots & u_N^M \\ u_0^{M-1} & u_1^{M-1} & \cdots & u_N^{M-1} \\ \vdots & \vdots & \ddots & \vdots \\ u_0^1 & u_1^1 & \cdots & u_N^1 \\ u_0^0 & u_1^0 & \cdots & u_N^0 \\ & & \underrightarrow{\ x\ } & \end{pmatrix}$$

The bottom row of this matrix $(u_0^0, u_1^0, \cdots, u_k^0)$ is the disturbance at time $t = 0$. You will select values for this row when you pull down the **Init** menu in the program. The first and last columns are determined by the boundary conditions. The task at hand is to find an algorithm that will allow us to generate any row in the matrix from a previous row or rows. As a first step, we write finite-difference approximations for the first derivatives of the function $u(x, t)$ as

$$\frac{\partial u(x,t)}{\partial t} = \frac{u_j^{n+1} - u_j^n}{\Delta t} \qquad \frac{\partial u(x,t)}{\partial x} = \frac{u_{j+1}^n - u_j^n}{\Delta x} \tag{3.27}$$

and the second derivatives as

$$\frac{\partial^2 u(x,t)}{\partial t^2} = \frac{u_j^{n+1} - 2u_j^n + u_j^{n-1}}{\Delta t^2} \qquad \frac{\partial^2 u(x,t)}{\partial x^2} = \frac{u_{j+1}^n - 2u_j^n + u_{j-1}^n}{\Delta x^2}. \tag{3.28}$$

The algorithms obtained when Eqs. 3.27 and 3.28 are substituted into the appropriate equation will have the same relationship to partial differential equations that the Euler algorithm has to ordinary differential equations; they are a useful starting point for further discussion. They may not be accurate, stable, or computationally efficient.

3.4.2   Klein-Gordon Equation; Numerical Method

The nonlinear Klein-Gordon equation,

$$\frac{\partial^2 u(x,t)}{\partial t^2} - \frac{\partial^2 u(x,t)}{\partial x^2} = F(u(x,t)), \tag{3.29}$$

is a good place to start our discussion, since the classical wave equation and the linear Klein-Gordon equation can both be treated as special cases of the function $F(u)$. Notice that the phase velocity is missing from the equation; this corresponds to a choice of units for which $v = 1$. Substituting Eq. 3.28 into Eq. 3.29, setting $r = \Delta t/\Delta x$, and rearranging terms, we obtain the following equation for the $(n + 1)^{-th}$ row of the U matrix in terms of the n^{-th} and $(n - 1)^{-th}$ rows:

$$u_j^{n+1} = -u_j^{n-1} + r^2(u_{j+1}^n + u_{j-1}^n) + 2(1 - r^2)u_j^n + \Delta t^2 F(u_j^n). \tag{3.30}$$

Ablowitz et al.[1] showed that this equation can be stabilized by setting $r = 1$ and using an average of the spatial coordinates for the function $F(u)$:

$$u_j^{n+1} = -u_j^{n-1} + u_{j+1}^n + u_{j-1}^n + \Delta t^2 F\left(\frac{u_{j+1}^n + u_{j-1}^n}{2}\right) \tag{3.31}$$

This choice of Δx and Δt not only reduces the number of computations, it also separates the U matrix into two non-interacting grids: $(j - n)$ mod $2 = 0$ and $(j - n)$ mod $2 = 1$; i.e., n and j have the same parity or opposite parity. This reduces the number of calculations by a factor of two, since only the even-even and odd-odd grid points need to be calculated. The numerical values for the current wave function and the previous wave function are stored in two global CUPS **DVectors** named **yVec** and **yPrevVec**, respectively. The corresponding x-values are stored in another **DVector**, **xVec**.

3.4.3 Classical Wave Equation; Numerical Method

We could use Eq. 3.31 to solve the classical wave equation numerically by setting $F(u) = 0$. Let's test this algorithm for a delta function. Set all but two of the values in **yVec** and **yPrevVec** equal to zero. Assume a value of u_j^n has just propagated during the third time step, $n = 3$, from the 15th to the 16th grid point. Then Eq. 3.31 gives the following result for the next time step:

$$(u_{15}^2 = 1, \ u_{16}^3 = 1) \ \overset{Eq.\ (3.31)}{\Longrightarrow} \ (u_{16}^3 = 1, u_{17}^4 = 1) \tag{3.32}$$

with all other values of $u_j^n = 0$. The algorithm tells us that the pulse continues to move to the right with velocity of 1. This result is *exact* since $f(x - t)$ is an analytical solution of the scaled wave equation. A similar argument can be made to show that $g(x + t)$ is the solution for a left-traveling wave. Since the equation is linear, these solutions can be added together to obtain a general solution. Rather than performing the extensive calculations given by Eq. 3.31 at every time step, we use the above results to simplify the calculation. We store the left- and right-going components in **yVec** and **yPrevVec** and then shift the elements of these vectors by one grid index every time step.

The above method must, however, be modified for a nonuniform medium. Since Eq. 3.31 requires that $\Delta x = \Delta t$, we need to change the grid-point spacing in order to reflect changes in the index of refraction, $\eta(x)$. A nonuniform grid is the price we pay for a simple algorithm. This new grid is called the optical grid, x_{og}, and has the property that the optical path difference, $\eta(x)\Delta x$, is the same between any two points. You will notice this nonuniform grid if you compare the spacing between data points on the screen inside and outside a segment.

Finally, we must incorporate the reflections caused by a change in index of refraction into the model. A reflection will couple the left-going and right-going waves and is included through the calculation of a reflection coefficient, r_j, at every grid point. The coefficient depends on the change in index of refraction which can be related to the change in grid spacing on the optical grid as follows:

$$r_j = \frac{\Delta x_{og}[j + 1] - \Delta x_{og}[j]}{\Delta x_{og}[j + 1] + \Delta x_{og}[j]}. \tag{3.33}$$

Notice that if the medium becomes more dense, then Δx_{og} decreases, and the reflection coefficient is negative, indicating a phase change.

3.4.4 Schrödinger Equation; Numerical Method

The Schrodinger equation can be discretized in a manner similar to the Klein-Gordon equation. In order to minimize the possibility of numbers becoming too large or too small for the computer, it is convenient to rewrite Eq. 3.18 using atomic units, i.e., $m = 1, \hbar = 1$. Variables are now on the order of unity and the Schrödinger equation becomes

$$i\frac{\partial u(x, t)}{\partial t} = -\frac{1}{2}\frac{\partial^2 u(x, t)}{\partial x^2} + V(x)u(x, t). \tag{3.34}$$

The propagation algorithm is obtained by substituting Eqs. 3.27 and 3.28 into this equation.

$$u_j^{n+1} = u_j^n + i\frac{\Delta t}{2\Delta x^2}\left(u_{j+1}^n - 2u_j^n + u_{j-1}^n\right) - i\Delta t V_j u_j^n. \tag{3.35}$$

Since the first derivative is approximated by a forward time step and the second derivative uses a difference centered on the current grid point, this approximation is often referred to as the forward time centered step (FTCS) algorithm. The values of u_j^n are, of course, complex numbers and will require two DVectors to store both real and imaginary values in the computer's memory. Although the FTCS algorithm is considered unstable, Pieter Visscher has shown that it can be stabilized if we alternate the calculation of the real and imaginary parts of the wave function and if the time step is sufficiently small.[17] Visscher's algorithm is reminiscent of the "leap frog" method for ordinary differential equations and has been implemented in **WAVE**. It can be written

$$r_j^{n+1} = r_j^{n-1} - \frac{\Delta t}{2\Delta x^2}\left(s_{j+1}^n - 2s_j^n + s_{j-1}^n\right) + \Delta t V_j s_j^n \tag{3.36}$$

$$s_j^{n+2} = s_j^n + \frac{\Delta t}{2\Delta x^2}\left(r_{j+1}^{n+1} - 2r_j^{n+1} + r_{j-1}^{n+1}\right) - \Delta t V_j r_j^{n+1} \tag{3.37}$$

where the real part of the wave function, $r(x,t)$, is evaluated at even time steps and the imaginary part of the wave function, $s(x,t)$, is evaluated at odd time steps. In addition, the time step, Δt, must satisfy the condition

$$\frac{-2}{\Delta t} \le V \le \frac{2}{\Delta t} - \frac{2}{\Delta x^2} \tag{3.38}$$

In order not to increase the number of variables in the program unnecessarily, we use the same DVectors that were created for the KG equation; the real part of u is stored in **yVec** and the imaginary part of u is stored in **yPrevVec**.

3.4.5   Diffusion Equation; Numerical Method

The last equation to be examined is the diffusion equation. The derivative terms are the same as the Schrödinger equation, so the FTCS algorithm becomes

$$u_j^{n+1} = u_j^n + K\frac{\Delta t}{\Delta x^2}\left(u_{j+1}^n - 2u_j^n + u_{j-1}^n\right). \tag{3.39}$$

This algorithm is stable only if $\Delta t < \Delta x^2/2K$, the so called Courant-Friedrichs-Lewy (CFL) condition. The more robust DuFort-Frankel algorithm[6] uses a time centered first derivative:

$$u_j^{n+1} = u_j^{n-1} + K\frac{2\Delta t}{\Delta x^2}\left(u_{j+1}^n - (u_j^{n+1} + u_j^{n-1}) + u_{j-1}^n\right). \tag{3.40}$$

The factor of 2 comes about because we are interpolating the rate of change over two time steps; the middle term in the second derivative expansion has also been replaced by a temporal average. Equation 3.40 must, of course, be solved for u_j^{n+1} before it can be implemented. It is stable for all values of Δt and is the algorithm used by the program. The algorithm requires the values of u at two times, t_n and t_{n-1}, in order to compute the next step. These values are stored in **yVec** and **yPrevVec**.

Although finite-difference methods are a respectable starting point for the solution of differential equations, they can become unstable for even modest values of Δt, and may therefore be computationally inefficient. The interested reader may want to study the techniques used by other CUPS authors such as the implicit Crank-Nicholson method used by Dan Styer in his QMTIME program and the relaxation method used by Bob Ehrlich and Jarek Tuszynski in their LAPLACE program, as well as more specialized books on numerical methods.[13]

3.4.6 Boundary Conditions

Three types of boundary conditions are implemented in the program: fixed, periodic, and absorbing.

Fixed Boundary

A **Fixed** boundary condition is the easiest to implement.

- For the classical wave equation the right- (left-) traveling wave is set equal to the negative of the left- (right-) traveling wave at the left (right) boundary.

- For all other equations, we set the function to zero at the boundaries.

As you will see in the exercises, these conditions result in a reflected wave for parabolic PDEs. For the diffusion equation, fixed boundaries will become "sinks" that eventually reduce the function to zero throughout the medium. The conservation of mass, $const = \int_a^b u(x,t)\,dx$, will therefore no longer be valid for this equation.

Periodic Boundary

Periodic boundaries are only slightly more difficult to implement. We require that the value of the function be calculated as if the ends were tied together. The wave at one endpoint is calculated as if it had the other endpoint as its nearest neighbor. This boundary condition is coded by using two "extra" grid points that have x-coordinates just outside the medium. These extra grid points are updated to the values at the opposite boundary after every time step but they are not displayed on the screen.

Absorbing Boundary

An **Absorbing** boundary is the most complicated.

- Since fixed boundaries for the diffusion equation already "absorb" or remove mass from the system, absorbing boundaries are redundant and have not been implemented for the diffusion equation.

- For the classical wave equation, we set the first grid point of the right-traveling wave or last grid point of the left-traveling wave equal to zero and let the propagation algorithm carry these zeros into the medium.

- Schrödinger wave functions are absorbed by adding $N/2$ extra grid points outside the medium, where N is the number of grid points displayed on the screen. These points are given a very small absorption coefficient, so that the wave is slowly attenuated after it crosses the boundary. In addition, the potential for these extra points is chosen to be that of the potential at the last point inside the medium, so that there is little discontinuity. This method works for all energies and typically results in an unwanted reflection of less than $1/N$.

- Absorbing boundary conditions have not been implemented for Klein-Gordon type equations, although a scheme similar to that employed for the Schrödinger equation would not be difficult to code.

3.5   Running WAVE

3.5.1   A Quick Tour of Wave

Load and start the program in the same way as other CUPS programs. The default configuration of a standing electromagnetic wave, shown in Figure 3.1, will appear on the screen. Purple and yellow represent the right- and left-traveling waves, while the white wave is their sum. The **F2 Run/Pause** hot key starts and stops the simulation. Certain hot keys (**F1-Help, F2-Pause, F3-Offset, F4-Pos dt, F5-Faster, F6-Slower**) are accessible while the simulation is running. When the simulation is paused, you may use the **F7-Reset** hot key to restore the initial configuration. The colored boxes in the lower left corner of the bottom graph are buttons that allow you to create reflectors (r), which reflect specified wave components: sources (s), which generate a disturbance; segments (n), which simulate refractive index, potential energy, or diffusivity depending on the type of simulation; and detectors (d), which monitor wave amplitude. Simply click once on the button with the label corresponding to the object you wish to create and click again on the wave graph at the position where you wish to insert the object.

To zoom in or out on the simulation graph, click the red "scale" button in the upper left corner of the graph and reset **xmin, xmax, ymin, and ymax**. Zooming is an option only while the simulation is paused. It changes the display without affecting the data being plotted.

The simulated time after n calculations, $n\Delta t$, is displayed on screen. Some of the simulations allow the user to reverse time by pressing the **F4-Pos dt/Neg dt** hot key. The direction of the time step can also be changed using the **Parameters | Time** menu. In addition to letting you "zero in" on interesting phenomena, reversing the time step is a useful check for the accuracy of the algorithm.

3.5.2 Controlling Execution

Depending on your computer (80386, 80486, or Pentium) you may need to adjust the speed of the animation. You can do this while the simulation is running by using the hot keys at the bottom of the screen or by using the **Preferences | Speed** menu. Although it was easy for us to decrease animation speed by adding a Pascal **PAUSE(t)** command, increasing the animation speed requires some type of trade-off. We have chosen to increase the animation speed by skipping screen redraws. This can produce a "jerky" wave, but because drawing frequency (as opposed to the time step, Δt), is varied, the underlying numerical algorithm is unchanged. Thus, the accuracy of the calculation is not affected by varying animation speed. **Caution**: Because analysis graphs record data only when the screen is redrawn, increasing the speed by performing multiple time-step calculations between screen redraws can result in distortion of analysis plots if the sampling frequency is small compared to the frequency of the simulated wave.

Adjusting the number of grid points under menu item **Parameters | Space Parameters** allows the user to trade accuracy for speed and is a good way to control animation. Setting **NumPts** should be done at the beginning of a session since it will clear the analysis graphs. Adjusting the time step is *not* a good way to control the speed. The classical wave, Klein-Gordon, sine-Gordon, phi-four, and double sine-Gordon calculations use an algorithm which automatically sets $\Delta t = \Delta x$; thus, the time step, Δt, is determined by the number of grid points,

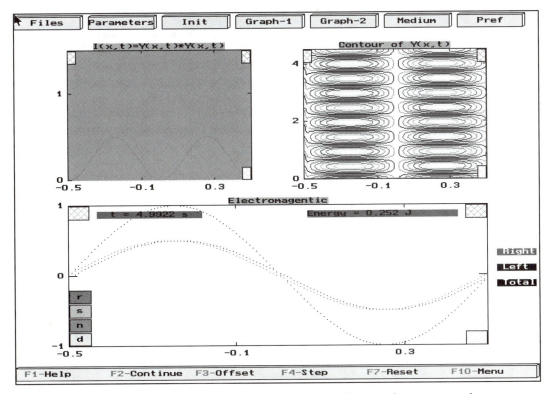

Figure 3.1: Opening screen with energy density and space-time contour plots.

which in turn specifies Δx. The Schrödinger and diffusion equations allow the user to set Δt independently of Δx by choosing the **Parameters | Time Parameters** option. Our algorithms are, however, unstable if Δt is too large. Whenever **NumPts** is changed, the program does the right thing. It automatically sets Δt to 80% of the maximum stable value. Since Δt is proportional to $1/$**NumPts**^{2} for both the Schrödinger and the diffusion equations, adjusting **NumPts** will have a dramatic effect on the calculation speed.

Data analysis can significantly affect the speed of program execution. Some graphs, such as the Fourier analysis of amplitude, $FFT\ y(x)$, require a great deal of calculation. Other graphs, such as the contour plot, spend most of their time just sitting in memory and collecting data and only once in a while perform the extensive calculations necessary for rendering. Expanding the wave graph to full screen will disable many analysis calculations and allow faster simulation. Time-consuming drawing will be eliminated but data will still be collected and stored by the analysis graphs. This data will be displayed when the wave graph is contracted.

Since control of the mouse is very CPU-intensive, disabling the mouse when the simulation is running will produce a noticeable speed enhancement. This can be done in the **Pref | Speed** menu.

Hint: Save the setup that provides the best compromise between smooth animation and accuracy for the Schrödinger and diffusion equations. Load these files as an alternate way to select the equation type.

3.5.3   Inspector Panels/Mouse-Accessible Objects

The WAVE program is built from encapsulated modules called *objects* which contain both data structures and the functions which act on the data. The user can access several of these objects directly through clicking or double-clicking the mouse, rather than through the menu. Double-clicking will generally bring up what we refer to as *inspector panels*, which are described below. If the CUPS utilities do not register a double click, you may need to adjust the double-click time using the **Files | Configuration** menu option. If the computer registers a single click, as opposed to a double click, an inspector panel will not appear— rather, the single click allows you to press a button or manually to adjust the position of a dragable reflector, detector, segment, or source by dragging the object to the desired position. Finally, if a single click cannot be associated with any object, the program will display the coordinates where the click occurred in the bottom left-hand corner of the graph.

- Reflectors (r)—Double-clicking a reflector brings up a **Reflector Attributes** panel which presents the following options:
    - Reflect the **left-going** component, **right-going** component, or **both** components
    - Set the **x-position** of the reflector
    - Specify the reflector as mouse-**drag**able or non-dragable
    - Get the reflector help screen
    - Delete the reflector

- Detectors (d)—Double-clicking the mouse on the desired detector brings up a **Detector Attributes** panel which displays the amplitude of the left-traveling component, the right-traveling component, and the total wave and presents the following options:

  - Set the **x-position** of the detector
  - Specify the detector as mouse-**drag**able or non-dragable
  - Get the detector help screen
  - Delete the detector

- Segments (n)—Double-clicking the mouse on a segment brings up a panel which allows you to modify the following parameters:

  - **x-position**—the left side of the segment is aligned with this position
  - **Width** of the segment
  - Refractive **index** in the case of an electromagnetic simulation **diffusivity** in a diffusion simulation **potential** in a Schrödinger equation simulation
  - **Gain**—allows for amplification of waves according to the formula

  $$E = E_0 e^{g\Delta x}, \tag{3.41}$$

  where g is the gain coefficient, E_0 is the incident electric field, and Δx is the distance traveled through the medium with gain coefficient g. Activating the **saturation** option under the **Medium** menu item replaces the user-specified gain coefficient with an "effective gain" given by

  $$g_{effective} = g_0(1 - E/10). \tag{3.42}$$

  - **Color** of segment
  - Specify the segment as mouse-**drag**able or non-dragable
  - Get the segment help screen
  - Delete the segment

- Sources (s)—Double-clicking the mouse on a source brings up a panel which allows the user to—

  - Set the function type as sine, Gaussian, pulse, step, modulated Gaussian (a sine wave multiplied by a Gaussian), white noise, or user-defined. To create a user-defined function, set the function type as **User Defined** and then select **Advanced Options | Define Function**.
  - Specify the source as mouse-**drag**able or non-dragable
  - **Delete** the source
  - **Exit** the source editor and implement the changes
  - **Adv**anced **Options** (only the second option is available for user-defined options)

    * Add **Noise** to the source. The **Amplitude** parameter specifies the maximum value for a random number that is added to the source term at each time step. The **Coherence Time** parameter, applicable only to sinusoidal sources, specifies the *average* time lag between phase shifts of random magnitude.

∗ Set direction of the wave generated by the source as bidirectional, right, or left
∗ Specify the source as **Periodic** or non-periodic (default is non-periodic), and set a time parameter (in seconds) which specifies the length of the source period. In addition to setting the frequency of periodic sources, the source period specifies the amount of time for which a non-periodic source will generate. Note that this option does not apply to sinusoidal or user-defined sources.

• Scaling—Clicking on the upper left (red) corner of any graph pulls up an inspector panel which enables the user to set the scale of the respective graph object. The user may set the scale by choosing the minimum and maximum values for each axis. Automatic scaling with respect to either or both axes is also available (options **X**, **XY**, and **Y**). Because data are stored in arrays which are independent of scale, changing the scale does not affect the data being plotted or result in a "loss" of data which lies off the visible portion of the graph.

• Expand/Contract Graph—The upper right (purple) corner of each graph acts as a toggle between a full-screen and smaller (default mode) graph.

• Attributes of Wave Graph—Clicking the lower right (blue) corner of the wave graph brings up a screen which allows the user to customize the display style of the wave graph. For example, in the case of an electromagnetic simulation, the user may—

– Select to show/hide the right-traveling wave, the left-traveling wave, and the sum of the two waves. In the default mode, all three of these waves are displayed.
– Specify the colors of the waves.
– **Offset Components**—displays different components of the wave on different axes. Deactivating this option places all waves on the x-axis.
– Connect the grid points which comprise the wave; the points will be connected only when the simulation is paused.

• Attributes of graphs 1 and 2 (the analysis graphs)—can be set by clicking on the lower right (blue) corner of the appropriate graph. All types of plots may be "frozen," so that they remain unchanged during continued simulation. Additional options are available for the following plots:

– **Space Integrated I(t)=Y*Y**—calculate a "spatial average" over a user-specified region in space. The value calculated represents one of the following, depending on the phenomenon being simulated:

∗ Total energy, electric energy, or magnetic energy of an electromagnetic wave (user specifies which energy from the attributes panel)
∗ Total energy or number of particles for a diffusion simulation
∗ Probability (of finding a particle within the region over which integration is carried out) for a Schrödinger simulation

– **Y(x,t)-Contour**—display color-coded legend, choose the palette, or change display to **Y(x,t)-3D**.

– **Y(x,t)-3D**—specify the Euler angles describing the orientation of the plot (this option allows you to rotate the plot), hide/show the "box" defined by the three axes of the plot, or change display to **Y(x,t)-Contour**.

– **Fourier Analysis of y(x)**—hide/show maximum values of the Fourier power spectrum; choose a basis set composed of sine terms or a basis set containing both sine and cosine terms (a basis set of sine terms is generally used to analyze simulations which fix the endpoints at $y = 0$).

– **Fourier Analysis of y(t)**—specify the number of points from which to construct the Fourier transform; choose a basis set composed of sine terms or a basis set containing both sine and cosine terms; hide/show maximum values of the Fourier power spectrum

– **Detector Readings**—display the left, right, or total wave in an electromagnetic simulation; display probability current in a Schrödinger simulation; calculate a **Time Average**. If the **Time Average** button is pressed, the temporal average of the selected wave component (right, left, or total; current) will be calculated over the specified time interval at the positions of the detectors

3.5.4 Menu Items

• **Files**

– **About CUPS**—Information about the CUPS project.

– **About the Program**—Information about the WAVE program.

– **Configuration**—Allows you to set a directory for temporary storage of overflow files, change the screen color, and check available memory.[*]

– **Open**—Loads an ASCII file containing a wave configuration from a drive. This option erases the current configuration and replaces it with the set-up specified by the file.

– **Merge**—Loads an ASCII file such that the file's wave configuration is *added* to the current screen's configuration. The merged file must be of the same equation type as the current set-up.

– **Save**—Saves the current wave configuration to a drive. Information pertaining to reflectors (r), detectors (d), segments (n), power sources (s), the time parameter, and the *selection of* analysis graphs will be saved. Note that the *data* displayed on analysis graphs will not be saved; these data must be regenerated when the file is loaded and run.

– **Save as**—Specify a filename and save the current configuration as above. Up to two lines of comments may be included in the file. These comments will be displayed when the file is loaded. You may change the target drive

---

[*]CUPS utilities buffer the screen to memory when doing a pop-up panel and restore the screen when the panel disappears. This speeds up the user interface since calculations are minimized, but the program must write to a temporary file on the disk if heap memory is not available. This can be slow without a cached hard drive and may cause a crash if the disk is full or write protected.

and directory by selecting the **[System]** option under the **[Pref]**erences menu item.

– **Exit**

- **Parameters**

  – **Space**—Set the number of grid points and the dimensions of the medium. For electromagnetic waves, this is the number of points in free space ($n = 1$). The number of optical grid points, which is the number of grid points you see on the screen, is adjusted automatically according to the local refractive index. For example, the density of grid points in a medium of $n = 2$ is twice as high as the density of grid points in free space. Larger numbers of grid points yield greater accuracy at the cost of a slower simulation. Note that changing the region of simulation or the number of grid points will destroy data, reset the graphs, and change the time step. In order to zoom in or out on the simulation without resetting it, the **Parameters | Scale Wave** option should be chosen or the upper left corner of the simulation graph may be clicked.

  – **Time**—Set the current simulation time and other time parameters. Setting the simulation time is useful when something interesting happens in the wave and you wish to reexamine a particular time interval. Change the time to 0 and run the simulation as desired. Resetting the graph **(F7)** will return the configuration to $t = 0$. The temporal direction of the simulation may be reversed by selecting or deselecting the **Positive dt** box. (While the program is running you can also use the **F4** hot key to toggle between positive and negative time steps.) You may adjust the value for the time step, Δt, used in diffusion and Schrödinger simulations, but Δt is automatically set such that $\frac{\Delta t}{\Delta x} = 1$ (for numerical stability) for electromagnetic, Klein-Gordon, sine-Gordon, double-sine Gordon, and phi-four equations. Note that a smaller time step will improve accuracy but decrease simulation speed. To adjust simulation speed *without* affecting numerical accuracy, use hot keys **F5-Slower** and **F6-Faster**, available while the program is running or select **Pref | Speed**.

  – **Scale Wave**, **Scale Graph1**, **Scale Graph2**— Adjust the scale of the wave graph, graph1, or graph2, respectively. These options are equivalent to double-clicking on the upper left (red) corners of the appropriate graphs.

  – **Electromagnetic, Klein-Gordon, sine-Gordon, phi-four, double sine-Gordon, Diffusion, Schrödinger**—Select which type of equation to simulate.

- **Init**—Specify the initial disturbance as random, sinusoidal, Gaussian, modulated Gaussian, pulse, symmetric pulse, modes, or a user-defined function.

- **Graph-1**, **Graph-2**—Disable the graph, clear the graph, or select one of the following analysis plots:

  – **Segment:n(x), V(x), or K(x)**—Displays the index of refraction, potential, or diffusivity in the medium depending on the equation type. Analytical values derived from **Medium | Parser** are added to segment values.

- **Space Integrated**—Displays the spatial integral of the modulus of amplitude as a function of time. This integral is proportional to energy associated with an electromagnetic wave simulation, probability ($\Psi * \Psi$) in the Schrödinger simulation, and the number of particles present in a diffusion simulation.

- **Y(x,t) Contour**—This graph displays isoclines representing points of equal amplitude as a function of position (on the x-axis) and time (on the y-axis).

- **Y(x,t) 3D**—Rather than using isoclines to represent amplitude on a two-dimensional graph, this option generates a three-dimensional depiction of amplitude as a function of position and time.

- **Fourier Analysis of Y(x)**—This Fourier power spectrum represents the transform of y-displacement as a function of x-position. This graph may slow down the simulation noticeably, because it is updated with each screen redraw. The red bars show the most recent transform, while the blue bars store maximum amplitudes obtained by each Fourier component.

- **Fourier Analysis of Y(t)**—This Fourier power spectrum represents the amplitude at a particular detector as a function of time. This graph is blank during an initial period of data collection.

- **Detector Readings**—Displays amplitudes at the detectors as a function of time.

- **Source Power**—Displays, as a function of time, the instantaneous power delivered by a source. Only available for electromagnetic simulations.

- **Intensity**—Depicts the modulus of amplitude as a function of position (on the x-axis) and time (the graph is continuously updated). This "intensity" represents a value which is proportional to energy in an electromagnetic simulation, probability ($\Psi * \Psi$) in a Schrödinger equation, and particle density in a diffusion simulation.

- **Medium**

  - **Boundary**—In the default (start-up) mode, both boundaries are **Fixed**, but **Periodic** and **Absorbing** boundaries are also available. When a wave encounters a **Fixed** boundary, it undergoes a "hard reflection" and a phase change of 180°. When the **Periodic** boundary condition is selected, waves traveling off the left-hand side of the screen re-enter on the right, and vice versa, creating a "wraparound" effect. The **Absorbing** boundary condition corresponds to boundaries which absorb the incident wave

  - **Saturation**—When the saturation option is activated, the gain of segments will be modified such that the total energy does not exceed 10:

$$g_{effective} = g_0(1 - E/10). \qquad (3.43)$$

  Note that if total energy, E, exceeds 10 when **Saturation** is activated, the resultant effective gain will be negative and the total energy will approach 10 asymptotically.

  - **Parse: n, V, or K**—enables user to describe refractive index, potential, or diffusivity as a function of position. The values generated by this function will be added to values specified by segments if segments are present.

– **Inspect Segments**—Enables user to inspect and modify each segment's attributes, one segment at a time. This option is especially useful if the configuration is such that one or more segments is "hidden" and cannot be accessed by double-clicking.

– **Remove Segments**—Removes all segments from the configuration

– **Inspect Sources**—Enables user to inspect and modify each source's attributes, one source at a time. This option is useful when the configuration is such that it is difficult to access sources through double-clicking.

– **Remove Sources**—Removes all sources from the configuration

- **Preferences**

  – **Display**—this option is equivalent to clicking the lower right corner of the wave graph (see section E.3).

  – **Files**—displays the following as editable fields:

    * **Drive and directory path for data files**—You may select a new path and/or another disk drive.

    * **Drive and directory path for overflow files**—You may select a new path and/or another disk drive.

  – **Speed**—Allows you to speed up the animation by increasing the number of time-step calculations carried out before the screen is redrawn, or to slow down the animation by inserting a Pascal **PAUSE(t)** command after every screen redraw. As with hot keys **F5-Faster** and **F6-Slower**, the accuracy of the underlying numerical algorithm is not affected by this option, because the time step, Δt, between calculations remains unchanged.

  – **Demo**—When **Demo Mode** is enabled, a demonstration simulation will evolve for the amount of time specified by the parameter **DemoTime**. The simulation will then reset itself and the demo will be run again. **Demo Mode** models whatever simulation— including graphs, detectors, reflectors, etc.— is present when enabled.

3.5.5   Hot Keys

At the bottom of the screen you will find a list of the hot keys and their functions. *Note:* The functions of the hot keys change when the simulation is running.

The following hot keys are available when the simulation is *not* running:

- **F1-Help**

- **F2-Continue** simulation

- **F3-Offset | No Offset**—Toggle between showing the wave components on separate/same axes, for the Schrödinger and classical wave equations.
  F3-Function | D/Dt—Toggle between showing the function or its derivative, for the diffusion, Klein-Gordon, sine-Gordon, double sine-Gordon, and phi-four wave equations.

- **F4-Step**—Proceed one time step, Δt, forward in the simulation.

- **F7-Reset**—Restore the initial configuration of $t = 0$.

- **F10-Menu**—Escape to the menu at the top of the screen.

The following hot keys are available when the simulation *is* running:

- **F1-Help**

- **F2-Pause**

- **F3**—Same as when the program is not running.

- **F4-Pos dT | Neg dT**—This key acts as a toggle between positive and negative time steps, causing the simulation to run forward or backward in time.

- **F5-Faster**—Increases the speed of the simulation *without affecting numerical accuracy* by increasing the number of time steps between each redrawing of the screen.

- **F6-Slower**—Decreases the speed of the simulation *without affecting numerical accuracy* by adding (or lengthening) a Pascal **PAUSE[t]** command between time steps.

3.6 Exercises

3.6.1 Visualization and Analysis Tools

Exercises in this subsection are designed to introduce important properties such as wavelength, frequency, eigenfunctions (i.e., modes) and dispersion. Readers who are familiar with these topics are urged to scan these exercises since the visualization and analysis tools that will be used for more advanced problems are also introduced.

3.1   **Classical Wave**
Load the WAVE program from DOS. The default values for the analysis graphs (energy density and contour) should be visible. Select a right-traveling Gaussian wave from the **Init** main menu. Run the program until the contour plot first appears and pause the program.

  a. What happens to the wave function at the right boundary? What happened to the energy distribution?
  b. Why does the density plot zig-zag and why do the colors alternate between red and blue? Expand the contour plot to full screen using the red expand button in the upper right-hand corner. Click on the blue button on the bottom right corner of the contour plot and change the attribute to 3D.
  c. Using the main menu, change the boundary condition to **Periodic** and rerun the experiment. (Click the **F7-Reset** hot key to reset the initial condition. Don't go to the main menu!) Explain the differences.

Note: You may need to adjust the program execution speed for subsequent exercises. The default setting may be too fast for the classical wave equation and too slow for the diffusion equation. Refer to the section on running the program for hints on how to adjust various parameters for optimum performance.

3.2   **Diffusion**

Change the equation type to diffusion under the **Parameters** menu, change the boundary condition back to **Fixed**, and place a Gaussian near the right boundary as the initial condition. Select the contour plot and **Space Integrated I(t)=y*y** as the analysis plots. Run the program until the contour plot first appears and pause the simulation.

a. What happens to the function at the right boundary? *Hint*: You may need to rescale the **Space Integrated I(t)=y*y** graph.

b. Using the main menu, change the boundary condition to **Periodic** and repeat the experiment. Explain the differences between a and b.

3.3   **Schrödinger**

Change the equation type to Schrödinger under **Preferences**. Select a Gaussian initial condition. Reselect the contour plot in one of the analysis plots and change the boundary condition back to **Fixed**. Run the program until the contour plot first appears and pause the program.

a. What happens to the wave function at the right boundary?

b. Change the boundary condition using the main menu to **Periodic** and repeat. Are there any differences?

c. Select the Gaussian initial condition but change the default conditions by setting the momentum (which happens to be equal to the wavenumber, k, since $\hbar = 1$) to zero. Run the program. What similarities do you see with the solution to the classical wave equation? To the diffusion equation?

3.4   **Klein-Gordon**

The behavior of a wave function can change markedly at different scales for wave functions with frequency-dependent phase velocity. In this exercise, the behavior of short wavelengths will be compared to that of longer wavelengths in a Klein-Gordon equation. Change the equation type to Klein-Gordon under **Parameters**. Resize the medium by setting the left boundary to -50 and the right boundary to 50 under **Parameters | Space Parameters**. Reselect the contour plot in one of the analysis plots and change the boundary condition back to **Fixed**.

a. Select a bidirectional Gaussian initial condition with a width of 1.5. Run the program until the contour plot first appears and pause the program. Notice the phenomenon of pulse propagation, as one might expect after having selected a bidirectional Gaussian. Although the pulses do not remain well-formed, disturbances propagate in both positive and negative directions, as before. Estimate the velocity of the disturbance from the slope of the outer contour isoclines.

b. Now select a Gaussian initial pulse with a width of 40.0. Run the program and note the differences from a. Clearly, the width of the wave function has a dramatic effect on the time evolution of the system modeled by the Klein-Gordon equation.

You should reset the space parameters to -0.5 and 0.5 after running the Klein-Gordon exercise. Rather than resetting the size of the medium, the boundary conditions, and the number of grid points to default values, we have chosen to change as few variables as possible when switching types. Unfortunately, the analysis graphs are very dependent on the equation type and must be reselected by the user whenever the equation type is changed.

Eigenfunctions

The previous exercises introduced some of WAVE's features. They suggest similarities and differences among the various equations; we will examine many of these similarities and differences in the following exercises. We begin by considering the *eigenfunctions* or *modes* for the four linear PDEs for a uniform medium of length L with fixed boundaries.

Various equations and eigenfunctions can be selected by first choosing an **EquationType** under **Parameters** followed by **Init I Modes** and then selecting the eigenfunction number. Modes 1 through 5 can be selected by checking the appropriate box. Modes that are not listed can be selected using the two input fields at the bottom of the screen. Select the **Fixed** boundary condition *before* you begin the following exercises; eigenfunctions are different for different boundary conditions.

3.5   **Fixed-Boundary Eigenfunctions**

a. Select **Parameters I Electromagnetic** and set the initial condition to mode 4. Create a detector by clicking on the blue button (labeled "d") in the lower left corner of the wave graph and then clicking inside the wave graph. Create two additional detectors. Run the program. Notice that all the detectors have the same frequency but different phases. Measure the frequency of this mode by holding down the mouse inside the detector graph and reading the time at two different maxima. Measure the wavelength by holding the mouse down in the wave graph and reading the separation between two maxima. Calculate the phase velocity, $v = \lambda f$. Place the system in another mode and repeat. Are the phase velocities the same?

b. Repeat a for the Schrödinger equation. Since the quantum detectors measure probability current, they cannot be used to measure phase oscillations. You can, however, follow a phase oscillation in the bottom graph by looking at the real (or the imaginary) part of the wave function. Examine modes 1, 4, and 16 and determine the frequency for each mode. Are the phase velocities the same?

c. Repeat a for the Klein-Gordon equation but set the ends of the medium to be at $\{-50, 50\}$. Try modes 1, 4, and 16. Are the phase velocities the same?

d. The functional relationship between wave number, $k = 2\pi/\lambda$ and the angular frequency, $\omega = 2\pi f$ for a harmonic wave is called a dispersion function. Determine the dispersion relationship, $\omega(k)$, for the equations in Table 3.2 that exhibit oscillatory behavior. A wave equation is said to be nondispersive if ω and k are proportional. Which of the equations in Table 3.2 are nondispersive? Which are dispersive?

e. Repeat a for the diffusion equation. Even though these solutions do not oscillate, there is a time associated with each mode, called the characteristic time, that determines how long it takes for the function to decay to $1/e$ of its initial value at a point. Determine the characteristic time for modes 1, 4, and 16.

3.6　**Periodic-Boundary Eigenfunctions**

Change the boundary to **Periodic** and examine the modes as in the previous exercise. It is important to note that these modes are *not* the same as for **Fixed** boundaries.

a. How do the eigenfunctions change when you switch to periodic boundary conditions? Try both positive and negative modes. Write the analytical form of the eigenfunctions for periodic boundaries.

b. Using **Periodic** boundary conditions with an **Electromagnetic** wave, place the system simultaneously in modes 4 and -4 and run the simulation. Is it possible to write the eigenfunctions for **Periodic** boundaries in terms of the functions listed in Table 3.2? Is the reverse possible?

3.7　**Beats**

The phenomena of beats is usually covered in introductory physics texts for the classical wave equation. You should now be familiar enough with the program to construct simulations that demonstrate the phenomenon of beats for other equations. Do you notice differences in the beat patterns when the simulation is run for the Schrödinger equation? The Klein-Gordon equation? *Hint:* Set the medium to $\{-50, 50\}$ for interesting effects with the Klein-Gordon equation.

Fourier Analysis

A real wave function can be decomposed into a sum of harmonic terms using the technique of Fourier analysis. With fixed boundaries this decomposition becomes

$$u(x,t) = \sum_{n=1}^{N} a_n(t) \sin(n\pi x/L) \tag{3.44}$$

Table 3.2　Eigenfunctions for various PDEs with fixed boundaries on $(0, L)$

Partial differential equation	Eigenfunction, $\phi_n(x,t)$	$\omega_n(k_n)$ where $k_n = n\pi/L$
$\dfrac{\partial^2 u(x,t)}{\partial t^2} - \dfrac{\partial^2 u(x,t)}{\partial x^2} = 0$	$N_n \cos \omega_n t \sin k_n x$	$\omega_n = k_n$
$\dfrac{\partial^2 u(x,t)}{\partial t^2} - \dfrac{\partial^2 u(x,t)}{\partial x^2} = u(x,t)$	$N_n \cos \omega_n t \sin k_n x$	$\omega_n = \sqrt{k_n^2 + 1}$
$\dfrac{\partial u(x,t)}{\partial t} - \dfrac{\partial^2 u(x,t)}{\partial x^2} = 0$	$N_n e^{-\omega_n t} \sin k_n x$	$\omega_n = k_n^2$
$i\dfrac{\partial u(x,t)}{\partial t} + \dfrac{\partial^2 u(x,t)}{\partial x^2} = 0$	$N_n e^{-i\omega_n t} \sin k_n x$	$\omega_n = k_n^2$

where N is the number of points on the space grid, L is once again the length of the medium, and n is referred to as the bin number. With periodic boundaries, both sin and cos terms are present in the decomposition.

$$u(x, t) = \sum_{n=1}^{N/2} a_n(t) \sin(2n\pi x/L) + b_n(t) \cos(2n\pi x/L)$$

$$= \sum_{n=1}^{N/2} c_n(t) \sin(2n\pi x/L + \delta_n) \tag{3.45}$$

For real $u(x, t)$ the Fourier analysis graph will display either a_n for fixed boundaries or $c_n = \sqrt{|a_n|^2 + |b_n|^2}$ for periodic boundaries. Expansion coefficients for a function defined on a uniform grid can be obtained very efficiently using a numerical technique called the fast Fourier transform, or FFT. It is discussed in detail in Chapter 2 and has been implemented in WAVE.

3.8   Real FFT

Load the WAVE program. Select **Graph-2 | Fourier Analysis of Y(x)** to enable the FFT of the function $u(x, t)$. Notice that the height of the red bar follows the oscillations of the wave when the program is running; the blue bar records the red bar's maximum value. Since the program default uses $N = 128$ points on the space grid, the bin scale on the abscissa may be too large. Zoom in on the FFT graph using the red button in the left-hand corner to display the scale inspector. Set **xMax** to 16. (Even though the abscissa measures the FFT bin number, we refer to the abscissa as "x" since the inspector is generic to all graph objects.)

a. The FFT graph *almost* shows the distribution of eigenfunctions that make up the E&M wave. If we write the solution of $u(x, t)$ in terms of the PDE eigenfunctions, ϕ_k,

$$u(x, t) = \sum_{n=1}^{N} \phi_n = \sum_{n=1}^{N} a_n \cos(2\pi nt) \sin(\pi nx/L) \tag{3.46}$$

and compare this expansion to Eq. 3.44, we see that the FFT coefficients have absorbed the time dependence:

$$a_n(t) = a_n \cos(2\pi nt). \tag{3.47}$$

Select a more complicated wave function and notice that the blue bar graph will show the eigenfunction distribution if the program runs through one oscillation of the lowest-frequency component.

b. Change the boundary condition to **Periodic**. Since the eigenfunctions depend on boundary conditions, you must now reselect the eigenfunction using **Init | Modes**. The FFT has changed, too. First, the red bar remains constant when the program is run. Why? Second, the number of bins on the abscissa is $N/2$. Why?

The choice of fixed or periodic boundary conditions selects between two different complete orthonormal sets of eigenfunctions. Both sin and cos terms are allowed for periodic boundaries while only sin terms are allowed for fixed boundaries. Fixed boundaries can, however, support half-integer wavelengths so the total number of functions is equal to the number of grid points in each set. We are interested in the frequency components and not in relative phases, so for periodic boundaries, the bar graph displays magnitudes of sin and cos of the same frequency in a single term of height $\sqrt{|a_n|^2 + |b_n|^2}$ and the number of bins in the FFT graph is reduced to $N/2$. A plot such as this, i.e., a plot of the magnitude of the Fourier components at a frequency, is called a *power spectrum* although we will continue to refer to our graphs as FFTs since this algorithm is the basis of our analysis.

3.9   Time Series FFT

It is also possible to perform a FFT analysis of time series data collected by a detector. You must first create a detector and then select **Graph-1 | Fourier Analysis- Y(t)**. Select **Graph-2 | Detector Readings** as the second graph and run the program. The analysis graph will appear after the detector has collected 256 data points and will again show the power spectrum, $\sqrt{|a_n|^2 + |b_n|^2}$. Remember that analysis graphs only collect data during a screen redraw so the time between data points is $\Delta t \times Speed$.

a. Calculate the total time interval during which the data was collected, T. Bin 1 contains the amplitude at $f_0 = 1/T$. Bin 2 contains the amplitude at the $2 \times f_0$, etc. Find the bin number with the maximum amplitude and calculate its frequency.

b. Is the time series FFT located in a single bin? What happens if the length of the time series is extended using the blue attributes button?

c. How does the time resolution (i.e., the animation speed) effect the time series FFT?

3.10   Complex Eigenfunctions

The solution to the Schrödinger equation can also be expanded in a Fourier series if the expansion coefficients, $a_k(t)$, are complex. Examination of the eigenfunctions shows that the expansion coefficients are phasors

$$a_n(t) = |a_n|e^{i\omega_n t} \tag{3.48}$$

with angular frequency $\omega_n \propto n^2$. Since the magnitude of any phasor is independent of time, we immediately obtain the eigenfunction distribution by plotting $|a_n(t)|$ as red bars in the analysis graph.

a. Set the **EquationType** to **Schrödinger** and the boundary to **Fixed**. Select modes 1 and 2 with amplitudes equal to 0.5. Select the FFT of $y(x)$ last so that the scale is properly set. Run the program. Notice that the particle probability oscillates between the left-hand and right-hand sides of the box. The bar graph produced by the FFT is, however, constant.

b. For a dramatic example of a constant FFT but a wildly varying wave function, try an initial Gaussian with width of 0.1 and average momentum (i.e., wave number) of 50. What you are seeing is called dispersion. Each mode has a different phase velocity, so the modes making up a particular function, in this case a Gaussian, drift out of phase producing distortion of the original shape.

c. Select mode 4 and notice the single FFT bin. Change the boundary to **Periodic**. Notice that there are now two nonzero bins and that the scale of the FFT graph has both positive and negative bin numbers. Explain the FFT.

d. Select mode 4 while the boundary is set to **Periodic** and then switch the boundary condition to **Fixed**. Explain the FFT.

The student of quantum mechanics should realize that each bin in the FFT analysis graph represents the probability amplitude of a momentum (kinetic energy) eigenstate when periodic (fixed) boundary conditions are selected.

3.11   Coherent States

An interesting case of a variable FFT and showing the correspondence between classical physics and quantum mechanics will be explored in this exercise. Enter $10 * x * x$ into the text field of the **Medium | Parser** input screen to create a simple harmonic oscillator (SHO) potential of $V(x) = 10^5 x^2$ in atomic units. Select a Gaussian initial wave function having zero average momentum, i.e., $k = 0$, and offset from the equilibrium point. Set the boundary condition to periodic in order to show momentum eigenstates. Run the program with analysis set to FFT of $y(x)$ and contour plots. Notice and explain the following:

a. The wave function remains a Gaussian as it propagates. Its FFT is also Gaussian.

b. The oscillations of the center of the wave function and the FFT are $\pi/2$ out of phase.

c. Try very narrow and very wide initial Gaussians. What do you observe about the width? The initial width that will propagate without changing shape is called a coherent state. Find this width. *Hint: Try the width of the ground state.*

Any Schrödinger wavefunction can be expanded in terms of SHO eigenfunctions. An interesting property of these eigenfunctions is that their energy eigenvalues, E_n, are equally spaced. The phase oscillations induced by the $e^{iE_n t/\hbar}$ factor in the time-dependent Schrödinger equation eigenfunctions are therefore harmonic and any initial wavefunction will reoccur after one phase oscillation of the ground state. Investigate this behavior of the following two wavefunctions using parser input under **Init | User Defined Function**.

d. A square pulse, $h(x - 0.3) - h(x - 0.2)$. The Heaviside step function, $h(x)$, is recognized by the parser.

e. A SHO wavefunction that has been offset from equilibrium. Enter the n-*th* SHO wavefunction offset from equilibrium, $\phi_n(x - 0.25)$. The SHO wavefunctions can be found in most quantum mechanics texts.

Green's Functions and Propagators

Any complete set of functions, not just eigenfunctions, can be used to construct a solution to a linear PDE if we know how these functions evolve in time. Dirac delta functions are a perfectly acceptable alternative to the trigonometric eigenfunctions that were studied previously. The time evolution of these delta functions is called a Green (or Green's) function.

We can approximate a Dirac delta function using a narrow Gaussian if we are careful not to make the width, a, smaller than the grid spacing in the simulation:

$$\delta(x) = \lim_{a \to 0} \frac{e^{-(x/a)^2}}{a\sqrt{\pi}}. \tag{3.49}$$

Figure 3.2 shows how an initial sawtooth disturbance is decomposed into delta functions.

3.12   Green's Function Visualization

Determine how a narrow delta function located in the center of the medium evolves in time under the action of the classical wave, diffusion, and Schrödinger equations. Select the equation type and then select contour plot. Use **Pref I Speed** to slow down the the display and to obtain good time resolution. Stop the simulation after the first rendering of the contours.

a. Can you guess the analytic form of the function seen on the contour plot?
b. Continue to run the program until the disturbance reaches the walls. Can you still guess the analytical form?
c. Place the initial Gaussian off-center. Do any of the contour plots just shift when the initial position changes or does the character of the contour plot change? Can you guess the analytical form even with arbitrary initial position for any of the PDEs?

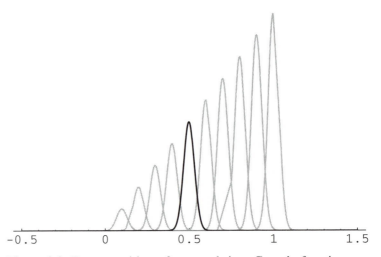

Figure 3.2: Decomposition of sawtooth into Green's functions.

You have found the Green's function for a particular PDE if you were able to answer the last question.

Clearly Green's functions exist for all of the linear PDEs in Table 3.1; you just displayed them as contour plots in the previous exercise! In practice, it is often difficult to find an analytic expression for these functions even if they are known to exist. Green's functions depend on the position of the delta function, x_0, at initial time, t_0, and must predict the disturbance at any other point in the medium, x, at any subsequent time, t. The disturbance will, of course, only depend on the elapsed time, $t - t_0$:

$$Green's\ function = G(x, t, x_0, t_0) = G(x, t - t_0, x_0, 0). \tag{3.50}$$

The Green's function can now be used to calculate the wavefunction at any subsequent time if the initial disturbance, $u_0(x) = u(x, 0)$, is known.

$$u(x, t) = \int_{\bar{x}=a}^{b} u_0(\bar{x}) G(x, t, \bar{x}, 0) \, d\bar{x}. \tag{3.51}$$

Since the Green's function for time-dependent PDEs allows one to calculate the wavefunction at any time, $u(x, t)$, using the initial conditions, $u_0(x)$, these functions are sometimes referred to as propagators.

3.13 Maxwell Reciprocity

Maxwell reciprocity states that the response at x_1 due to a delta function at x_0 is equal to the response at x_0 due to a delta function at x_1— provided that the elapsed times are the same:

$$G(x_1, t_1, x_0, t_0) = G(x_0, t_1, x_1, t_0). \tag{3.52}$$

Use detector graphs to demonstrate this symmetry. Set up the system with a narrow initial Gaussian and a detector at another point. Run the simulation and then freeze the graph using the blue attribute button on the detector graph. Create another detector graph and reverse the positions of the delta function and the detectors. Notice that the two detector graphs are identical.

A formal representation of Green's functions is not hard to derive. Substituting the definition of the Fourier coefficients, a_n, into an eigenfunction expansion of a solution, we obtain

$$u(x, t) = \sum_{n=1}^{\infty} a_n \phi_n(x, t) = \sum_{n=1}^{\infty} \left[\int_{0}^{L} u_0(\bar{x}) \phi_n(\bar{x}, 0) \, d\bar{x} \right] \phi_n(x, t). \tag{3.53}$$

If we interchange the order of the summation and integration and compare the result to Eq. 3.51, we obtain the Green's function as a sum over eigenfunctions:

$$u(x, t) = \int_{0}^{L} u_0(\bar{x}) \left[\sum_{n=1}^{\infty} \phi_n(x, t) \phi_n(\bar{x}, 0) \right] d\bar{x}, \tag{3.54}$$

$$G(x, t, \bar{x}, 0) = \sum_{n=1}^{\infty} \phi_n(x, t) \phi_n(\bar{x}, 0). \tag{3.55}$$

3.14   Finite Space Functions

Use a function plotting package such as MathCAD, Maple, or Mathematica to plot the eigenfunction expansion of the Green's function for the diffusion equation for different values of $\bar{x}$.

$$G(x, t, \bar{x}, 0)$$

$$\approx \sum_{n=1}^{N} (2/L) \sin(k_n x) \sin(k_n \bar{x}) e^{-\kappa(k_n)^2 t} \text{ where } k_n = n\pi/L \tag{3.56}$$

Does the expansion work best for short or long times? Close to or far from the boundary? Compare your result to contour plots produced by WAVE.

3.15   Infinite Space Functions

The Green's function expansion simplifies considerably if the medium is very large so that boundaries can be ignored. The summation in Eq. 3.55 becomes an integral which can be evaluated for the wave equation, Schrödinger, and diffusion equations. Compare the short time evolution of a narrow Gaussian located near the center of the medium to the functions listed in Table 3.3.

Superposition and Interference

Restart the program or select **Electromagnetic** and **Fixed** boundaries if the default conditions have been changed. Select an energy density data analysis graph, $I(x, t)$, and use the **F3-Offset** hot key to separate the left, right, and total wave components.

3.16   Energy

Use **Init I Gaussian** to select an initial right-traveling wave centered at $x = -0.25$ with an amplitude of 0.5 and a width of 0.1.

   a. Integrate the square of the initial wavefunction, $\int_a^b u(x, 0)^2 \, dx$. What is the relationship of this integral to the **Energy** reading displayed in the blue stripe in the wave graph?

   b. Run the simulation until the peak of the Gaussian coincides with the right boundary. What is the value of the integral now?

   c. Use **Init** to create a second left-traveling Gaussian centered at $x = +0.25$ with an amplitude of 0.5 and a width of 0.1. Check the add to existing wave button and run the simulation until the two pulses overlap. Does the energy reading change when the two waves overlap? Does the integral change?

Table 3.3   Green's functions for various PDEs with boundaries at infinity

Partial differential equation	Green's function, $G(x, t, \bar{x}, 0)$
Wave equation	$\delta(x - t)/2 + \delta(x + t)/2$
Diffusion equation	$(2\pi t)^{-\frac{1}{2}} e^{-(x-\bar{x})^2/2t}$
Schrödinger equation	$(2\pi i t)^{-\frac{1}{2}} e^{i(x-\bar{x})^2/2t}$

Be sure to consider the magnetic field for a traveling wave when you answer the previous questions. If the electric field is positive, a right-traveling wave will have a magnetic field pointing out of the screen while a left traveling wave will have a magnetic field pointing into the screen. Note also that the phase shift at a boundary is different for magnetic and electric fields.

3.17 Standing Wave Energy

Use **Init | Modes** and select mode 4. Select two energy distribution data analysis graphs, $I(x, t)$. Click the blue attributes button on the left analysis graph and select **Electric**; click the blue attributes button on the right analysis graphs and select **Magnetic**. Run the program and explain the behavior of these two graphs. What is the relationship between the nodes of the standing wave and the zeros in the analysis graphs?

3.18 Quantum Interference

Select **Schrödinger** and **Fixed** and an energy distribution data analysis graph, $I(x, t)$. Select an initial Gaussian wavefunction centered at $x = -0.25$ with an amplitude of 0.5, a width of 0.1, and a momentum of $+50$. Add to this a second Gaussian centered at $x = 0.25$ with an amplitude of 0.5, a width of 0.1, and a momentum of -50. Collide these two Gaussian pulses but pause the simulation when the pulses overlap. What do you observe in the analysis graph? Use a mouse-down to measure the separation of the zeros in the analysis graph. How is this separation related to the momentum?

3.19 Multiple Sources

Set both boundary conditions to absorbing. Create three overlapping sinusoidal sources at $x = -0.25$ with the following frequencies and amplitudes respectively: $f_i = (1, 3, 5)$ and $a_i = (1, 0.333, 0.2)$. Set the **NoDrag** button to true, so that the sources will not inadvertently be moved. The output of the sources should approximate a square wave, since we have selected the first three nonzero Fourier coefficients of this function. Verify this.

3.20 Coherence Time

No natural sinusoidal source is a perfect monochromatic wave; it must have a starting and a stopping time. If we assume that a 1 Hz source is at maximum amplitude at some time, $t = 0$, then it might be safe to assume it will be at maximum at $t = 2$ or $t = 3$ seconds. But is it safe to assume a maximum at 100 seconds? The length of time that we can make a prediction about the phase of a sinusoidal source is called the *coherence time*. The coherence time has been implemented for sinusoidal sources in the **Adv Options** submenu of the **Source Inspector**.

a. Create a medium with absorbing boundaries containing a detector at $x = 0.4$ and a sinusoidal source with the following properties:

  * Position: -0.4
  * Frequency: 10 Hz
  * Amp: 0.5
  * No Drag: TRUE
  * Noise: TRUE
  * Amp Noise: 0.0

∗ Coherence Time: 1.0

∗ Direction: Right

Select a detector analysis plot and an FFT of the detector readings, FFT-Y(t). Use the blue attributes button to set the number of points in the FFT to 1024. Set the display speed to 1 for maximum time resolution. Run the program. Notice the abrupt phase change approximately every second. The FFT is no longer a single peak but a distribution of frequencies. Using paper and pencil, calculate the Fourier transform of a sine wave of finite duration.

b. Create 8 additional sources identical to and overlapping the first source. These sources are initially in phase and their amplitudes add to 4.5. They slowly drift out of phase with each other and after about 1 second their phases are random. Estimate the r.m.s. amplitude and standard deviation of N sources with random phases and compare to the readings on the detector plot.

Resonance

3.21   E&M Sources

Set the equation type to electromagnetic and select **Init | Zero** to clear the wavefunction. Click on the red source button in the function display and create a source by clicking in the middle of the display. Exit the inspector panel and run the simulation with the default parameters. Notice that a right-traveling Gaussian pulse is produced at a time of 0.5 seconds and that this pulse passes through the source after reflecting from the fixed boundary. Double-click on the source to bring up the inspector panel again. You can now adjust the source parameters using the sliders and buttons in the panel. Clicking the **Adv Options** button displays a second panel with additional source properties. Forcing the wavefunction to follow the source has a dramatic effect on the simulation, since it effectively adds a boundary condition to the medium. Run the program to see the effect. The **Periodic** switch allows the user to select a repetition rate for Gaussian and most other pulses. **Periodic** has no effect on sinusoidal or user-defined functions. **Coherence Time**, on the other hand, affects only sinusoidal sources.

3.22   Power

Consider the standard introductory physics demonstration consisting of a string fixed at one end with a tuning fork at the other. The forced end produces a wave which is reflected from the fixed end, then reflected again from the forced end. This gives rise to resonance effects; the phase relationship between the traveling waves and the source determines the amplitude of the resonance. Start this exercise by selecting **Parameters | Electromagnetic** and **Init | Zero** and choosing **Fixed** boundary conditions. Create a source and set the following properties:

• Function Type: Sine
• No Drag: TRUE
• Amp: 0.5
• Freq 1.0

- Pos: -0.5
- (Advanced Options) Direction: Right
- (Advanced Options) Track Source: TRUE

Select space integrated, $E(t) = \int_a^b I(x,t)\,dx$, and source power, $P(t) = dE(t)/dt$, analysis plots.

   a. Run the program and explain the behavior of the two analysis graphs. What is the phase relationship between the reflected, i.e, left-traveling, wave and the source? Change the source frequency to 1.1 and repeat after hitting **F7** to reset. What happens now?

   b. In order to simulate the "real" system you will need to add some damping. Create a segment and edit its properties as follows:

- Position: -0.50
- Width: 1.0
- Index: 1.0
- Gain: -0.1
- Color: Whatever looks good!
- Able to Drag: FALSE

Run the program until a steady state is reached and record the energy for frequencies from 0.5 to 1.5 in increments of 0.1. Plot these energies and predict the steady-state energy of the system as a function of frequency. Note that the source end will not be a node even though it acts like a fixed boundary for the traveling wave. Can you predict how the position of the node closest to the source will vary with frequency?

Steps and Barriers

The close connection between quantum mechanics and optics will be explored in the following exercises. For many types of phenomena a variation in potential energy is shown to be equivalent to a variation in index of refraction.

  3.23 **Electromagnetic Step**

Load WAVE from DOS and set both boundaries to **Absorbing**. First create a segment by clicking on the green segment button in the bottom left-hand corner and then clicking inside the wave graph. Now double-click inside the segment to bring up the inspector, and edit the parameters as follows:

- X Position $= 0.0$
- Width of Segment $= 0.5$
- Index of Refraction $= 1.5$
- Gain $= 0$
- Color of Segment $= 2$
- Able to Drag $=$ FALSE (i.e., not checked)

Close the inspector by clicking OK. Use **Init | Gaussian** to create a right-going Gaussian pulse with the following attributes:

- Amplitude $= 1$
- Width $= 0.1$
- Position $= -0.4$

a. Calculate theoretical reflectance for a light wave of normal incidence on glass with $n = 1.5$.

b. Create detectors to monitor the reflected and transmitted wave and enable a detector analysis graph. How does reflectance of the peak compare to the prediction of theory? Note the inversion of the reflected pulse. Why is the pulse inside the medium narrower than the original pulse?

c. Measure the speed of the incident and transmitted pulses. Is this what you expected?

d. Change the initial condition so that the pulse is a left-going pulse starting inside the medium. How does reflectance of the peak compare to the prediction of theory? Is the reflected pulse inverted? Why does the transmitted pulse have a larger amplitude than the incident pulse?

e. Note the continuity of the electric field at the boundary between two media.

f. Calculate the **Time Average** of the square of the incident, transmitted, and reflected Gaussian pulses (you will find this option in the menu brought up by the attributes button of the detector analysis graph). Make sure you have the Sqr(Amp) radio button checked in the Average Inspector panel. How do these average values compare to the prediction of theory? *Hint:* Is the transmitted energy the square of the electric field?

3.24   Electromagnetic Barrier

For what frequencies (between 1 and 15 Hz) of normally incident light will reflection from a sliver of glass ($n = 1.5$) with thickness $d = \frac{1}{12}$ be maximized? For what frequencies will transmission be maximized? Check your answers through simulation and determine the reflectance at the maxima.

3.25   A One-Layer Antireflection Coating

Near the left edge of the wave graph, place a source producing a right-traveling sinusoidal wave with a frequency of 10 Hz. Put a glass segment ($n_{glass} = 1.5$) at $x = 0.0$, extending to the right boundary. Create a one-layer antireflection film on the left face of the glass segment. The thickness of the film should be $\frac{\lambda}{4}$, where λ is the wavelength of the incident wave *inside* the coating (adjusted for n_{film}).

a. Compare the simulated reflectance to the theoretical prediction given by

$$R = \left(\frac{n_{glass} - n_{film}}{n_{glass} + n_{film}} \right)^2 .$$

(3.57)

b. What is the reflectance of the wavelength, λ_0, for which the film was created? What is the reflectance for $\frac{2\lambda_0}{3}$, $\frac{4\lambda_0}{5}$, and $\frac{5\lambda_0}{4}$?

c. Change the antireflection medium to magnesium fluoride (n_{MgFl_2} = 1.35). Observe the resultant reflectances for λ_0, $\frac{2\lambda_0}{3}$, $\frac{4\lambda_0}{5}$, and $\frac{5\lambda_0}{4}$.

3.26   Quantum Step

Select the Schrödinger equation and create a segment with a potential energy of $-1.0 * 10^4$ in the region $(0.0, 0.5)$. Set the initial wave to be a Gaussian with the following conditions:

- Amplitude = 1
- Width = 0.3
- Position = −0.25
- Momentum k = 100

Create two detectors, one inside the segment and one outside the segment at $x = -0.15$. The detectors measure probability current and can be used to determine the tansmission and reflection coefficients. The boundaries should be set to **Absorbing**. Select a **Graph-2 | Detector Readings** analysis graph. Run the program.

a. Notice that the Gaussian pulse breaks apart into transmitted and reflected components reminiscent of the classical wave simulation. Notice also that the spatial frequency of phase oscillations increases as the potential energy decreases, in keeping with an increase in kinetic energy. Click the blue attributes button on the detector graphs and perform a time integral of the detector readings. Since the detectors read current, J,

$$J = \frac{1}{2i}\left(\Psi^* \frac{\partial \Psi}{\partial x} - \Psi \frac{\partial \Psi^*}{\partial x}\right) \tag{3.58}$$

the integrals represent the probability of a particle's passing a given detector.

b. Increase the potential step to $0.5 * 10^4$ by double-clicking on the segment and editing the inspector. Run the program and estimate the reflection and transmission coefficients from the pulse heights and widths. Notice that the phase wavelength decreases inside the higher-potential segment.

c. Increase the potential to $+1.0 * 10^4$. The pulse is almost entirely reflected by the step, although it does penetrate the forbidden region. Use the integral of the probability current to show this.

Answer (partial): Atomic units simplify many of the relationships between wave and particle properties. A particle with momentum $k_1 = 100$ in a zero-potential region has energy $E = K_1 = k_1^2/2 = 0.5 * 10^4$ and wavelength $\lambda_1 = 2\pi/k_1 = 0.063$. If this particle-wave propagates into a region where $V_2 = -1.0 * 10^4$, the kinetic energy increases to $K_2 = E - V_2 = 1.5 * 10^4$ and the momentum becomes $k_2 = \sqrt{2(E - V_2)} = 173$. Use the coordinate reader to check the wavelength: hold the mouse down inside the wave graph and move it between two phase maxima.

3.27   Quantum Antireflection Coating

Create a double potential step by adding a second segment of width d to the left of a potential step or $(V = +0.25 * 10^4)$. Set the potential and width of this new segment so that the following conditions are satisfied:

$$k_f = \sqrt{k_1 k_2} \qquad d = \lambda_f/4 = \pi/2k_f. \tag{3.59}$$

Run the program. Notice that the reflection of the quantum wave packet is almost zero. Use detectors and current integrals to determine the reflection and transmission probabilities.

Multilayer Films

Agreement with theory for the following exercises is very dependent on the number of grid points. A minimum of 512 grid points is recommended.

3.28   A Two-Layer Antireflection Coating

Model a two-layer antireflection film for light of wavelength λ_0 incident on dense barium crown glass for which $n = 1.61$. Use zirconium dioxide ($n = 2.1$) and cerium flouride ($n = 1.63$) to construct the two-layer film. The thickness of both layers should be $\frac{\lambda}{4}$, and the light should encounter the lower-index layer (cerium flouride) first. Compare the reflectance given by the transfer matrix of the system to the simulated reflectance for λ_0. Repeat the previous comparison for $\frac{2\lambda_0}{3}$, $\frac{4\lambda_0}{5}$, and $\frac{5\lambda_0}{4}$. How must the antireflection film be changed in order to achieve a reflectance for λ_0 which approaches zero? Remember that in order to achieve $R(\lambda_0) \rightarrow 0$, the following condition must be satisfied:[8]

$$\left(\frac{n_2}{n_1}\right)^2 = \frac{n_{glass}}{n_{air}}, \tag{3.60}$$

where n_1 and n_2 are the refractive indices of the first and second materials encountered by incident light.

3.29   A Three-Layer, Two-Wavelength Antireflection Coating

Construct and test a three-layer antireflection coating that is a perfect reflector at two wavelengths.

3.30   An Eight-Layer Reflector

Create a high-reflectance film consisting of eight alternating layers of $\frac{\lambda}{4}$ thickness of materials with high and low refractive indices. Let the material of high refractive index be zinc sulfide ($n_h = 2.3$) and the material of low refractive index be magnesium fluoride ($n_l = 1.35$). Compare simulated reflectance with theoretical reflectance, R, given by

$$R = \left(\frac{(\frac{n_h}{n_l})^{2N} - 1}{(\frac{n_h}{n_l})^{2N} + 1}\right)^2, \tag{3.61}$$

where N is the number of two-layer stacks in the coating.

3.31   Using Thin Films to Model a Fabry-Perot Interference Filter

Construct a Fabry-Perot interference filter by coating both sides of a res-onating cavity with a three-layer partially reflecting film of alternating low- and high- refractive index materials (see Fig. 3.3). What are the simulated and theoretical reflectances of this seven-layer Fabry-Perot interference filter? How does the filter respond to a noisy source?

3.32   Modeling a Transmission Filter

Create a multilayer filter which allows for complete transmission of λ_0 but reflects other wavelengths (see Fig. 3.4). If you have access to a mathematical computation package, plot the reflectance as a function of wave number, $R(k)$, for filters of 1, 2, 3, and 4 layers. How does the $R(k)$ vary with the number of layers used in the filter?

3.33   A Beam-Splitter

Modify the filter configuration of the previous exercise such that the spaces between layers are occupied by segments. Adjust the refractive index of these segments such that the transmittance and reflectance of λ_0 are both approximately $\frac{1}{2}$.

3.34   Finite Pulse Width

Load the Fabry-Perot interference filter stored in the file FABRY.WAV and make the following changes. Double-click on the source and set the source function to user defined. Enter the following user defined function by clicking the **Advanced Options** button and entering the following formula in the parser text field:

$$e^{-(t-2)^2} \sin 6\pi t. \tag{3.62}$$

Select detector graphs for both analysis graphs and use the blue attribute buttons to set one detector graph to show the right going wave and the other to show the left going wave. Run the simulation. What is the width of the pulse transmitted by the filter? The effect you see is very important when designing optics for femtosecond laser systems. Almost any optical component will have an effect on pulse shape.

Resonators

3.35   Electromagnetic Decay

Place an eight-layer reflector (see Exercise 30 and Fig. 3.5) at $x = 0$ and

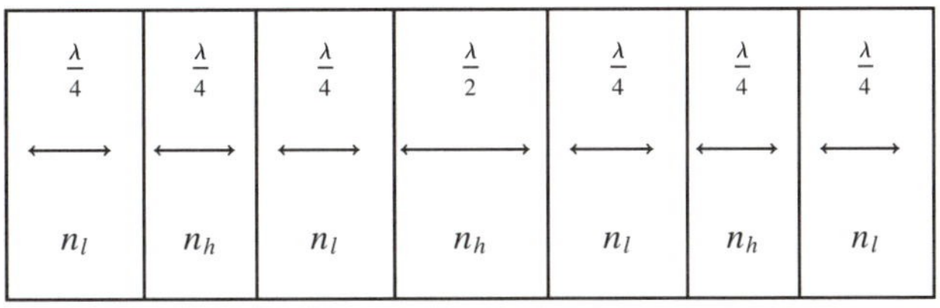

Figure 3.3: A multilayer Fabry-Perot interference filter (n_l=low refractive index, n_h=high refractive index).

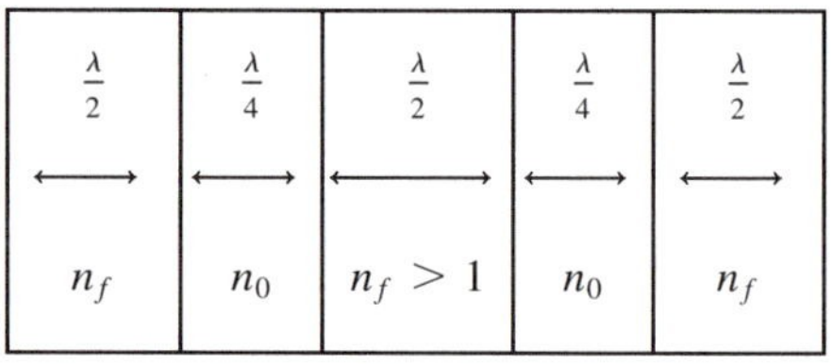

Figure 3.4: A multilayer transmittance filter, $n_f > n_0$.

set the left boundary to fixed and the right boundary to absorbing. Set the initial wavefunction so that there are an integral number of wavelengths, $n\lambda_0$, between the left boundary and the reflector. Plot the total energy as a function of time and explain the relationship of this plot to the quality factor, Q.

3.36   Lasers

Use the previous exercise to create a laser. Do this by creating a segment that has gain and fills the cavity. Adjust the gain such that the total amplification in one pass exceeds the transmission loss through the reflector, where

$$amplification = e^{2gL} \tag{3.63}$$

and L is the length of the cavity. Remember that g can be positive or negative and that g will be reduced as the total energy in the system approaches 10 if you have saturation turned on by using **Pref**.

   a. Calculate the threshold gain and see what happens to the initial sine wave when you create a segment filling the region$\{-0.5, 0\}$ with index $n = 1.0$ and threshold gain. Try gain values above and below the threshold value.

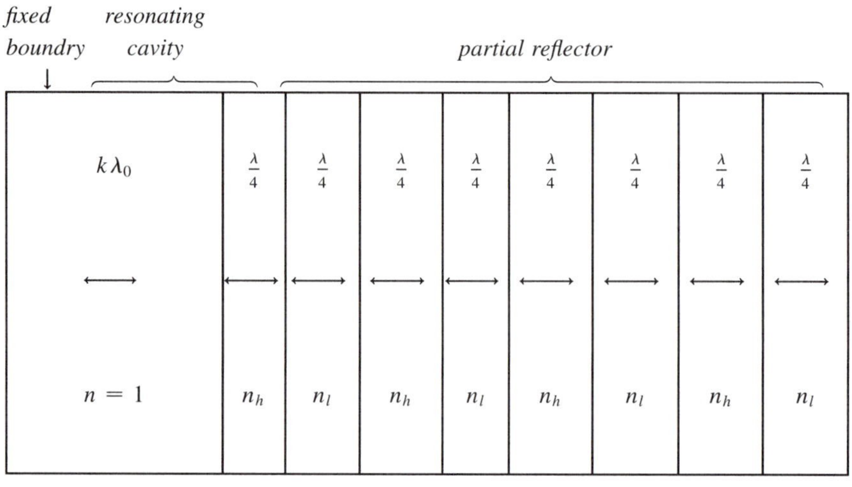

Figure 3.5: Laser configuration for exercise 36. n_l=low index of refraction; n_h=high index of refraction.

b. Initialize the laser configuration with a Gaussian wavepacket inside the resonating cavity. Set the gain to 1.2. Is lasing established at a single frequency? Why or why not? Look at an FFT of the output detector reading, $FFT\ y(t)$, with the number of FFT points set to 1024 for maximum frequency resolution. Actual laser cavities will exhibit similar phenomena. They will lase simultaneously in a number of modes.

3.37 Quantum Decay

Create a standing wave to the left of a narrow (but high) potential barrier. Plot the total probability as a function of time. Although the decay looks like it might be exponential, and many authors use this system as a model for nuclear decay, it is not. It is a power law for long times. Both tunneling and packet spreading are occurring![11,12]

WKB Approximation

Inhomogeneous media, i.e., media with variable index of refraction or potential, also have eigenfunctions, although the calculation of these functions is more difficult than for the stepwise homogeneous case considered in the previous section. Analytical solutions can be found only for very special cases; thus, in general, one must employ numerical techniques such as the shooting method described by Ian Johnston elsewhere in this series. An approximation, the Wentzel-Kramer-Brillouin (WKB) method, can often be employed when the properties of the medium vary slowly on a scale comparable to the wavelength. This approximation can be used for both the classical wave and Schrödinger equations and is a very useful guide for estimating solutions to these equations. It is based on the idea that we can estimate the wavelength based upon the properties of the medium. The wavelength in the classical-wave or Schrödinger equations will be shorter if the medium is optically more dense or the kinetic energy is larger, respectively.

WKB wavefunctions can be obtained by assuming that the wavefunctions are oscillatory— these are, after all, hyperbolic PDEs— and that the total phase of the oscillations in a medium with fixed (periodic) boundaries must be half-integer (integer) multiples of 2π. Since the wavenumber, $k = 2\pi/\lambda$, gives the number of radians per meter at x, we require that a medium with fixed boundaries at a, b satisfy

$$\int_a^b k(x)\,dx = n\pi. \tag{3.64}$$

The unnormalized WKB wavefunction is

$$\phi(x) = \frac{1}{\sqrt{k(x)}} \sin \int_a^x k(x')\,dx'. \tag{3.65}$$

Clearly, the function $\phi(x)$ will be zero when $x = a$, since the integral is zero, and again $\phi(x) = 0$ when $x = b$, since the integral will be $n\pi$ by the previous condition, Eq. 3.64.

3.38 Classical WKB

The classical WKB approximation can be credited to Lord Rayleigh who used it in 1912 to analyze the propagation of an optical disturbance in a nonuniform medium. It is discussed on page 310 of Elmore and Heald.[3] Rayleigh's approximation is equivalent to assuming that the wave number is proportional to the index of refraction $k(x) = k_0 \eta(x)$ where $k_0 = 2\pi/\lambda_{vacuum}$. Substituting this expression into Eq. 3.64, we see that there are an infinite number of values of k_0 (or λ_{vacuum}) that satisfy the condition. This is actually a trivial eigenvalue equation,

$$k_0 \int_a^b \eta(x)\, dx = n\pi, \tag{3.66}$$

that requires the evaluation of a single integral. Once the values of k_0 are known, the WKB wavefunctions can be calculated using Eq. 3.65.

   a. Enter the following linear position-dependent index of refraction using **Medium | Parser**.

$$\eta(x) = x. \tag{3.67}$$

     Since the parser adds its values to the values already existing in the medium, the actual index of refraction will be $\eta(x) = 1 + x$. Select mode 32 and run the program. Notice that the antinodes do not have the same amplitude and are not uniformly spaced. Are the overtones harmonics?

   b. Using pencil and paper, evaluate Eqs. 3.64 and 3.65 and compare with the results obtained from a.

   c. Construct an index of refraction that does not satisfy the condition that changes occur slowly on the scale of one wavelength and demonstrate that the WKB wave functions constructed by WAVE are not eigenfunctions.

   d. *(Advanced)* The exact analytical solution to the classical wave equation with a linear variation in the medium can be written in terms of Airy functions.[5] Compare this analytical solution to the WKB solution obtained in b.

3.39   Quantum WKB

Obtaining WKB solutions to the Schrödinger equation is more difficult than for the classical case. The wave number is not a linear function of potential, $k(x) = \sqrt{2(E - V(x))}$, and the wavefunction changes from an oscillatory to an exponential function if the potential is larger than the energy E. In addition, the WKB approximation fails in the transition region where $E - V(x)$ changes sign because the wavelength becomes large and encompasses appreciable variation in potential. We have therefore chosen to limit our WKB calculation to systems where the wavefunction is oscillatory, i.e., the classically-allowed regions where $E - V(x) > 0$, and have set the wavefunction equal to zero in the classically forbidden region. The eigenvalue equation for E_n, Eq. 3.64, is first solved by numerical methods:

$$\int_a^b \sqrt{2(E_n - V(x))}\, dx = n\pi \tag{3.68}$$

and each eigenvalue is substituted into Eq. 3.65 to obtain the wave-function.

    a. Select the Schrödinger equation and fixed boundary conditions. Enter the following potential using **Medium | Parser**:

$$V(x) = 0.04 * x, \tag{3.69}$$

and then select an initial condition of mode 6. Notice that WKB eigenfunctions are very similar to the WKB solutions of the classical wave equation from the previous exercise.

    b. Change the potential to a ramp:

$$V(x) = 0.01 * (h(x) * 5 * x - h(x - 0.2) * (5 * x - 1)), \tag{3.70}$$

and study the time evolution of the following WKB eigenfunctions, $m = 1, 2, ...7$. For what modes is the WKB approximation appropriate? Why is there a spike in the wavefunction for mode 3?

Nonlinear Klein-Gordon Equations

Constant solutions to nonlinear Klein-Gordon (NKG) equations are easy to identify. Not only is $u(x, t) = 0$ a solution to all three model equations, but any root of the function $F(u) = 0$ will also satisfy Eq. 3.19. For example, the phi-four equation has three constant solutions, $u(x, t) = 0$ and $u(x, t) = \pm 1$, while the sine-Gordon equation has an infinite number of solutions, $u(x, t) = n\pi$. Although these solutions are trivial and uninteresting in and of themselves, they often represent stable states of the system. Any soliton would likely approach such a state as $x \to \pm\infty$. If a system has multiple stable states, it is easy to imagine a solution that takes the system from one stable state to another; such a solution is called a kink if $u(x, t)$ increases as x increases and an anti-kink if $u(x, t)$ decreases. Unfortunately, a kink (or anti-kink) is topologically impossible for our system since all solutions must match the boundary conditions. Kink/anti-kink pairs are possible and will become the basis for some interesting simulations. (Fig. 3.6)[1]

    Here are a few hints on how to use some of the program's features to complete the following exercises. Try to construct your solutions one piece at a time. You can then edit the two parameters in the formula that determine the initial position and velocity of the soliton. The parser input screen allows you to add new functions to the current wavefunction. It is also possible to merge two WAVE data files together using **Files | Merge**. All existing settings are preserved; only the wavefunction is changed. This will be particularly useful for multi-soliton collisions. Finally, be sure to set the medium to $[-20, 20]$, since all parameters in the exercises have been selected to give good results for this setting.

    3.40   **Nonlinearity**

    Set the size of the medium to $-20, 20$, using **Parameters | Space Parameters** and increase the number of grid points to 512. Set the boundary to fixed and select the sine-Gordon equation.

    a. Place the system in mode 2 with an amplitude of 0.1. The system will warn you that it has chosen Klein-Gordon rather than sine-Gordon

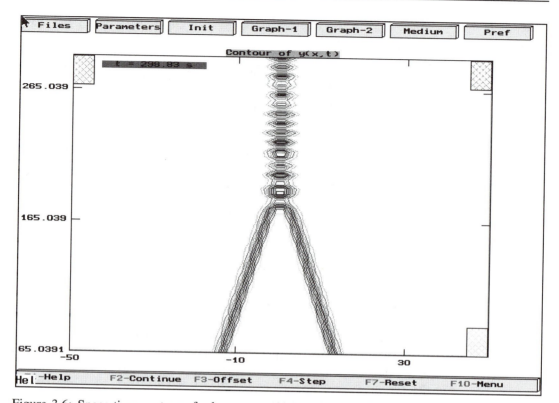

Figure 3.6: Space-time contour of a low energy kink-antikink collision in the phi-four equation.

eigenfunctions for reasons that will be obvious after you complete the exercises. Run the simulation with a contour analysis graph. Notice that the system acts pretty much like the a Klein-Gordon eigenfunction. This is not suprising since $\lim_{x \to 0} \sin u \approx u$ for low amplitude waves.

b. Select mode 2 again but this time set the amplitude to 10. Run the program. You should, of course, use the red scale button to rescale the wave graph. What has happened to the oscillation? Clearly the nonlinearity has had an effect since the new wavefunction is not 100 times the old wavefunction.

c. Repeat b with periodic boundary conditions. Be sure and reselect **Init | Modes** after you change the boundary.

3.41   Sine-Gordon Equation

A sine-Gordon kink can be written

$$u(x, t) = 4 \arctan e^{\gamma(x - vt)}, \tag{3.71}$$

where $\gamma^2 = (1 - v^2)^{-1}$ and v is the velocity of the kink.

a. Use pencil and paper to show that Eq. 3.71 satisfies the sine-Gordon equation.

b. Set fixed boundaries at $[-20, 20]$ as before. Use **Init | User Defined Function** to set the initial state to be a kink (Eq. 3.71), with $v = 0.8$.

It is important that you do *not* set t equal to zero. The program will evaluate the function at $t = 0$ to determine **yVec** and again at $t = -\Delta t$ to determine **yPrevVec**. Does the kink propagate as expected? Measure the velocity of the kink. Notice that the boundary introduces a discontinuity.

c. Write an analytic expression for an anti-kink.

d. Initialize the following kink/anti-kink state:

$$u(x, t) = 4 \arctan e^{2.7777(x - 0.8t + 8)} - 4 \arctan e^{2.7777(x + 0.8t - 8)}. \tag{3.72}$$

It is convenient to type each arctan function into a separate text field since the two text fields on the parser input screen are concatenated. Run the simulation with fixed boundaries.

e. Run the simulation with periodic boundaries. Notice that each collision moves the system down the ladder of stable states! We actually plot $u(x, t) \bmod 8\pi$ to keep from having to rescale the screen.

f. Initialize the following kink/anti-kink state with periodic boundary conditions.

$$u(x, t) = 4 \arctan e^{1.3333(x + 0.5t - 4)} - 4 \arctan e^{2.7777(x + 0.8t - 12)} \tag{3.73}$$

Notice how the anti-kinks retain their shape as they pass through each other, even though the system is both dispersive and nonlinear.

3.42   Double Kink

We have seen that two sine-Gordon solitons with different velocities pass through each other. What happens if you superimpose two solitons on top of each other with the same speed? The resulting function will carry the system though a change of 4π. Will such a state propagate without changing shape?

3.43   Can You Find a Soliton?

Try a variety of different initial large amplitude wavefunctions and notice how quickly nonlinearity destroys their shape. Clearly solitons are very special solutions.

3.44   Phi-Four Equation

The interaction inherent in the phi-four equation produces velocity-dependent effects. Although soliton-like behavior is observed at high velocities, low velocities can result in complex resonance phenomena and even in bound states. We will again construct a topologically allowed solution by combining kink ($+$ sign) and anti-kink ($-$ sign) solitons.

$$u(x, t) = \pm \tanh \gamma (x - vt)/\sqrt{2} \quad \text{where } \gamma^2 = (1 - v^2)^{-1} \tag{3.74}$$

Select the phi-four equation and create a grid with 512 points and periodic boundaries at $x = \pm 20$ as before. Place a detector at $x = 0$ and select a detector analysis plot along with a contour analysis plot. *Note:* The u^3 term in the phi-four simulation makes the simulation susceptible to numerical overflow—and a possible program crash. The chances of this happening are minimized if you select a large number of grid points.

   a. Create a kink at $x = -10$ with $v = 0.9$ and an anti-kink at $x = +10$ with $v = -0.9$. Make sure that the function you input begins and ends in the stable $x = -1$ state and that the kink carries the wavefunction into the stable $x = +1$ state. The boundaries should, of course, be periodic so that the ends do not add an unstable $x = 0$ solution. Run the simulation and observe the soliton-like behavior— at least for the first collision.

   b. Create a kink at $x = -10$ with $v = 0.1$ and an anti-kink at $x = +10$ with $v = -0.1$. Run the simulation and describe the motion. The long-lived bound state is called a breather/bion. Notice the small traveling waves. They represent radiation leaving the interaction region.

   c. Create a kink at $x = -10$ with $v = 0.224$ and an anti-kink at $x = +10$ with $v = -0.224$ and run the simulation. Notice the long collision time. There are a number of such resonances between $v = 0.193$ and 0.258.

3.45   Double Sine-Gordon Equation

The double sine-Gordon equation is similar to the sine-Gordon equation in that it has an infinite number of stable states but these states are separated by 4π rather than 2π. A kink solution exists and is given by

$$u = 4\arctan e^{\sqrt{5}/2(x-vt)\gamma+\Lambda} + 4\arctan e^{\sqrt{5}/2(x-vt)\gamma-\Lambda}, \tag{3.75}$$

where $\Lambda = \ln(\sqrt{5} + 2) \approx 2.118034$ and $\gamma^2 = (1 - v^2)^{-1}$. You should use the **F3** hot key to toggle to a *derivative* plot of the wavefunction before running this simulation in order to highlight properties of the soliton.

   a. Plot Eq. 3.75 and its derivative. How does the separation of the two peaks in the derivative depend on the velocity parameter?

   b. Set periodic boundaries and create a double sine-Gordon kink at $x = -10$ with $v = +0.15$ and anti-kink at $x = +10$ with $v = -0.15$. An easy way to do this is to create a kink with the parser and exit the parser screen. You can then re-enter the parser, create an anti-kink, and add it to the existing kink using the add to existing wave option. Another way would be to create kink and anti-kink files and use **Files | Merge** to combine these functions. Run the program. Are the kinks true solitons? Do they pass through each other or do they interact?

   c. Modify part a by changing the value of Λ to 5. What happens now? The solutions are no longer solitons but they still propagate a disturbance.

3.46   Multiple Soliton Collision

Our previous NKG exercises have involved kink/anti-kink collisions, but other interesting combinations are topologically possible. Set up multiple-soliton collisions and determine if such a collision destroys the kinks. What happens if a high-velocity kink collides with a breather/bion solution in the phi-four equation?

Numerical Effects

The accuracy and precision of a numerical simulation are highly dependent on the resolution of the space-time grid. In order to keep the users from shooting themselves in the foot, we recalculate these parameters whenever a significant change occurs in the system. For example, the time step is reset whenever the equation type is changed or a segment is created, destroyed, or moved. The default value of the time step, Δt, is equal to 80% of the maximum stable value for the diffusion and Schrödinger equations and is equal to Δx for Klein-Gordon type equations.

3.47   Accuracy

Load the Fabry-Perot transmission filter data file FABRY.WAV. Compare reflectances obtained in these simulations with the values obtained with the default number of points, 128, and with theoretical predictions. Remember that for the classical wave equation

$$\Delta x = \Delta t = (x_{right} - x_{left})/N, \qquad (3.76)$$

where x_{left} and x_{right} are the left and right boundaries of the medium and N is the number of points. Describe the relationship between number of points and accuracy of the simulation. Is the algorithm first-order or second-order in the number of points? For example, what happens to the error when the number of points is cut in half? How does computation time depend on the number of points?

3.48   Instability

Select the Schrödinger equation as the equation type and set the initial condition to be a Gaussian. Set the display refresh to 1 using **Pref | Speed**. Use **Parameters | Time Parameters** and increase Δt by a factor of 2 above the default value. Use the **F4** hot key to single-step the algorithm. Clearly our algorithm has become unstable.

3.49   Spurious Waveforms

A change in the behavior of a solution is much harder to detect than the catastrophic failure of the Schrödinger equation for large Δt noted in the previous exercise. A good way to check for these types of errors is to increase the number of grid points or to decrease the time step (or both!) and see if the character of the solution changes. Use default values for the number of grid points for the following exercises.

   a. The DuFort-Frankel algorithm used to solve the diffusion equation is stable for all values of Δt; it may not be accurate. Examine the behavior of the diffusion algorithm with large time steps. Change the initial disturbance to a Gaussian pulse after selecting the equation type. Use **Parameters | Time Parameters** and increase Δt by a factor of 100 above the default value. Run the simulation and notice the non-physical behavior of the simulation.

   b. Reexamine the Schrödinger equation solution to the simple harmonic oscillator. Instead of $10 * x * x$, enter $100 * x * x$ into the text field of the **Medium I Parser** input screen. Select a Gaussian initial wave-function having no average momentum, i.e., $k = 0$, but offset from the equilibrium point. Set the boundary condition to periodic in order to show momentum eigenstates. Run the program with analysis set to FFT of $y(x)$ and contour plots. Run the simulation and notice the spurious wavefunction. How must the system parameters be changed to model the dynamics of this system accurately?

Program Modification

3.50   **Diffusion Current**

It is often possible to change the physics being displayed on the screen by a slight modification of a key procedure. One such procedure is the **CalculateGridPointDensity** function. Since the value returned by this function is displayed in the **Graph 1 I I(x,t)** data analysis graph, changing this function will allow the user to display other dynamical properties that can be calculated at a grid point.

a. Use the finite-difference approximation to write an expression for the current density, $J(x, t)$, associated with the diffusion equation

$$J(x,t) = -K\frac{\partial u(x,t)}{\partial x} \tag{3.77}$$

in terms of the current values of the wavefunction which are represented on the grid by u_j^n.

b. Rewrite the **PASCAL** code for the **CalculateGridPointDensity** function in the **WAV-GBL.PAS** unit so that it returns the probability current when EQUATIONTYPE=DIFFUSION. Do this by adding another match to the **CASE** statement. The current will now be shown whenever you select the **Graph 1 I I(x,t)** data analysis option. Are you able to predict the observed distributions when the initial condition is a mode? A Gaussian?

3.51   **Energy on a String**

The classical wave equation (Eq. 3.3), models the transmission of waves on a string as well as E&M waves. For small displacements we can write the total energy density for this system as

$$w(x,t) = \frac{\mu}{2}\left(\frac{\partial u(x,t)}{\partial t}\right)^2 + \frac{T}{2}\left(\frac{\partial u(x,t)}{\partial x}\right)^2, \tag{3.78}$$

where μ is the mass per unit length and T is the tension.

a. One difference between these two systems is that the energy density exhibits frequency-dependence in addition to the usual amplitude-squared dependence. Find an analytical expression for the instantaneous and average energy per unit length for a harmonic wave, $A\cos(kx - \omega t)$, traveling on a string.

b. Find an expression for the energy density at the n-th grid point. Use the forward finite difference approximation for the first derivatives of u_j^n. Assume that $T = 1$ and that $\mu = 1$.

c. Rewrite the **PASCAL** code for the **CalculateGridPointDensity** function in the **WAV-GBL.PAS** unit using this new expression. The **I(x,t)** option under **Graph** will now plot this density. Are you able to predict the observed distributions when the initial condition is a mode? A Gaussian?

3.52   Quantum Probability Current

Although not as widely used as probability density, the current density

$$J(x, t) = \frac{i}{2}\left(\Psi(x, t)\frac{\partial \Psi * (x, t)}{\partial x} - \Psi * (x, t)\frac{\partial \Psi(x, t)}{\partial x}\right) \qquad (3.79)$$

is an equally useful quantity for many types of problems.

a. Find an analytical expression for the probability current density for right- and left-traveling de Broglie waves, $A\exp[i(kx \pm \omega t)]$.

b. Find an expression for the probability current at the $n-th$ grid point. Use the forward finite difference approximation for the first derivative of u_j^n.

c. Rewrite the **PASCAL** code for the **CalculateGridPointDensity** function in the **WAV-GBL.PAS** unit so that it returns the probability current when EQUATIONTYPE=SCHRODINGER. Examine the current distribution for various initial conditions using the **Graph 1 | I(x,t)** option. Are you able to predict the observed distributions when the initial condition is a mode? A Gaussian?

d. Set up the Schrödinger equation with absorbing boundaries. Build two barriers and fill the region between these barriers with a standing wave using the parser. Is the probability current what you expect?

3.53   Advection Equation

It order to understand the numerical solution of Eq. 3.3 it is often helpful to begin by considering a simpler but related equation, the advection equation:

$$\frac{\partial u(x, t)}{\partial t} = -v\frac{\partial u(x, t)}{\partial x}. \qquad (3.80)$$

This equation has many properties in common with the classical wave equation, but it has the very interesting property that it supports propagation of a function $u(x, t)$ in only a single direction.[6]

a. Show that the solution of the classical wave equation is given by a linear combination of $u(x - vt)$ and $u(x + vt)$, while the solution of the advection equation is only given by $u(x - vt)$.

b. What change must be made to Eq. 3.80 to admit solutions of the form $u(x + vt)$?

c. Use finite-difference methods to solve this equation. Find an expression for the u_j^{n+1} in terms of previous grid points. Rewrite the propagation code for the Procedure **CWStep** using your new algorithm. Is your algorithm stable?

Acknowledgments

At various stages of this project we were helped considerably by students and colleagues at Davidson College. We would particularly like to thank Laurence S. Cain for editing the manuscript and testing the software and Roger Freedman for help interpreting the various nonlinear Klein-Gordon equations.

References

1. Ablowitz, M.J., Kruskal, M.D., and Ladik, J.F. SIAM Journal of Applied Mathematics 36:999, 1979.

2. Corson, P., Lorain, D. *Electromagnetic Fields and Waves*. San Francisco: Freeman Press, 1988.

3. Elmore, W., Heald, M. *Physics of Waves*. New York: Dover, 1985.

4. Fowles, G.R. *Introduction to Modern Optics*. New York: Holt, Reinhart and Winston, 1968.

5. Fulcher, L.P. American Journal of Physics, **53:**730, 1985.

6. Garcia, A. *Numerical Methods for Physics*. Englewood Cliffs, NJ: Prentice Hall, 1994.

7. Georgi, H. *The Physics of Waves*. Englewood Cliffs, NJ: Prentice Hall, 1993.

8. Hecht, E., Zajac, A. *Optics*. Reading, MA: Addison-Wesley, 1974, p. 315.

9. Kragh, H. American Journal of Physics. **52:**1024, 1989.

10. Main, I.G. *Vibrations and Waves in Physics*, 3rd ed. Cambridge, England: Cambridge University Press, 1993.

11. Massmann, H. American Journal of Physics **53:**679, 1985.

12. Onley, D., Kumar, A. American Journal of Physics **59:**562, 1991.

13. Press, W.H., Flannery, B.P., Teukolsky, S.A., Vetterling, W.T. *Numerical Recipes: The Art of Scientific Computing*. Cambridge, England: Cambridge University Press, 1986, p. 632.

14. Shiff, L.I. *Quantum Mechanics*, 3rd ed. New York: McGraw-Hill, 1968.

15. Visscher, P.B. *Fields and Electrodynamics*. New York: John Wiley and Sons, 1988.

16. Visscher, P.B. Computers in Physics **3:**42, 1989.

17. Visscher, P.B. Computers in Physics **5:**596, 1991.

4

Interference and Diffraction

Robin A. Giles

> Rays of light are very small bodies emitted from shining substances.
> Query 29: *The Third Book of Opticks*
> Sir Isaac Newton, 1704

> So it arises that around each particle there is made a wave of which that particle is the centre.
> *Traité de la Lumieré*
> Christian Huyghens, 1690

4.1 Introduction

One of the interesting pieces in the physics puzzle is found in the work of Issac Newton. This giant worked in many areas of physics and the subject of optics did not escape his attention. Of note was that he considered light to be particulate in nature. Later Hooke, Young, and Huyghens favored a wave approach. We note that we now consider both approaches to have their validity. But not so in the 18th century. The phenomenon of interference and diffraction proved particularly easy to understand in terms of waves but somewhat intractable for Newton with his adherence to a particle approach.

The program DIFFRACT allows you to look at some fundamental wave behaviors through a study of both Fresnel and Fraunhofer diffraction, and some of the interference and coherence effects. In particular, you will be able to study diffraction phenomena associated with and a slit or set of slits and a point or set of points using the Huyghens construction. You can also use a method developed by Cornu—the Cornu Spiral—to examine diffraction from one or two slits or from one or two obstacles. You can study Fraunhofer diffraction with a single slit or set of slits, a rectangular aperture and a circular aperture. Finally, you can study partial coherence and the topic of visibility in interference and diffraction observations. In the latter example you will be able to study the Michelson stellar interferometer and model the measurement of the separation distance in a double star and measure the diameter of single stars.

4.2   The Distinction Between Interference and Diffraction

When a point source of light casts a shadow of an object onto a screen the laws of geometrical optics predict that a sharp distinction should be seen between the region of shadow and the region that is fully illuminated.Close examination shows this not to be the case and the edge of the shadow is not distinct. Diffraction concerns this deviation from the principles of geometrical optics. Light does not always travel in straight lines because an object can cause the light to bend. Although this was first observed by Grimaldi some one hundred years earlier, it was Christian Huyghens who is most closely identified with the first formal study of the phenomena—through Huyghens' principle. Huyghens proposed in 1678 that every vibrating point on a wave front became the source of a new disturbance. These secondary disturbances together form an envelope with the same properties as the originating wave front and the wave propagates, as in Figure 4.1.

If an obstacle is placed in the line of propagation, light far from the obstacle continues unaffected, but close to the edge the secondary wavelets originating from the wave near this edge find no counterpart to join with from the region in the shadow. The wave front turns around the edge as a result. In a different experiment over a century later, Thomas Young examined the intensity of light falling on a screen when light from a single source passed through two parallel slits. He observed a fringe pattern too, which, using Huyghens' principle, he attributed to the superposition of waves emanating from the two slits which act as two sources radiating in phase.

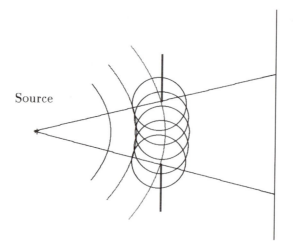

Figure 4.1: Huyghens' principle.

The first example is defined as *diffraction* and the second as *interference.* In the former example the fringe pattern is the result of the superposition of waves coming from various parts of the *same* wave front, whereas in the latter case the fringe pattern results from superposition of waves emanating from two *different* wave fronts. Such a clear distinction is rarely found because in the case of the inteference pattern found with two slits there are also diffraction effects resulting from the superposition of waves coming from the different sections making up the wave from one slit. Like the phenomenon itself the distinction is blurred.

4.3   The Classes of Diffraction

When the source of radiation is at a finite distance from the diffracting obstacle the wave front is curved as it reaches the diffracting obstacle and the diffraction effects are classified as *Fresnel diffraction.*

If the source and screen are at infinity or if a lens is used to produce parallel light at the obstacle and a lens is used to focus the resultant radiation onto a screen at a finite distance from the obstacle, then the diffraction is classed as *Fraunhofer diffraction.*

Fresnel diffraction effects are easier to observe and were historically the first to be examined. The mathematical basis is somewhat more complex, however, and the contributions of Fresnel are of great importance. The fact that light traveled in straight lines proved to be one of the greatest difficulties in the early development of the wave theory of light. Newton postulated that light was corpuscular in nature; straight-line propagation naturally followed and he accounted for the diffraction effects as the result of the deflection of the corpuscles by the edge.

There are problems with Huyghens' ideas where the wave front acts a source of a secondary spherical wave. On this model there is a backward convergent wave as well as forward diverging wave. This backward wave does not exist. Huyghens introduced an *obliquity factor* to remove it. This factor provides for a variation in amplitude as function of angle and

$$\text{obliquity factor} = 1 + \cos \theta, \tag{4.1}$$

where θ is defined as in Figure 4.2.

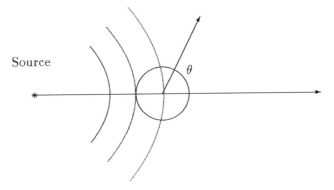

Figure 4.2: Obliquity factor.

4.4   Fresnel Diffraction

4.4.1   Huyghens' Principle

Two different examples are examined. The intensity at a point in a plane is evaluated for a slit and a set of slits and for a single point source or a set of identical point sources placed along a line. The mathematical analysis is essentially the same, except that in the case of the slit the obliquity factor is taken into account, whereas for a point source the radiation is spherically symmetric.

The wave equation has a general solution describing spherical waves[2]:

$$y(r, t) = \frac{A}{r} \cos (\omega t - kr + \phi), \tag{4.2}$$

where y is the amplitude of the wave, k is the wave number, r is the position, ω is the angular frequency, t is the time and ϕ the phase angle.

For N such spherical waves of the same frequency but different amplitudes and phases combining at a point

$$y_{total}(r, t) = \sum_{j=1}^{N} \frac{A_j}{r} \cos (\omega t - kr + \phi_j), \tag{4.3}$$

and the intensity at a point is given by

$$I(r, t) = \left(\sum_{j=1}^{N} \frac{A_j}{r} \cos \phi_j \right)^2 + \left(\sum_{j=1}^{N} \frac{A_j}{r} \sin \phi_j \right)^2. \tag{4.4}$$

4.4.2   Cornu Spiral

Consider light from a *line* source falling on a straight edge, where the source is a finite distance from the edge, so that the wave front reaching the edge is curved.[7] In this case we will be dealing with cylindrical wave fronts. In Figure 4.3, *CD* represents the wavefront as it reaches the edge. The source is placed at *O*. The illumination at a point P_1 on the screen will come from points like M_1 on the unscreened wavefront. If d is a small section of the wavefront at *A*, the edge itself, and if the amplitude at *A* is proportional to $\cos \omega t$, then the amplitude at P_1 due to ds at *A* will be $\cos (\omega t - kb) ds$. An element at M_1 on the wavefront away from *A* will provide an amplitude $\cos [\omega t - k(b + \delta)] ds$, where the distance from M_1 on the wavefront to P_1 is $(b + \delta)$. These amplitude functions should include a distance factor b and $(b + \delta)$, but we can equate these two factors if the screen is at distance that is large compared to δ, which will be the case here.

The total amplitude at P_1 will then be the sum of such contributions along the complete wavefront,

$$A_{total} = \int \cos (\omega t - k(b + \delta)) ds, \tag{4.5}$$

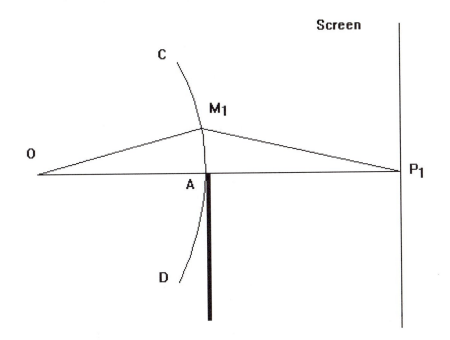

Figure 4.3: Diffraction at an edge.

and the resultant intensity (Eq. 4.4),

$$I = \left(\int \cos k\delta\, ds \right)^2 + \left(\int \sin k\delta\, ds \right)^2. \tag{4.6}$$

For points not too distant from the edge

$$\delta = \frac{s^2(a + b)}{2ab}, \tag{4.7}$$

where s is the length of the arc along the wavefront from A to M_1, and a and b are the distances from source to edge and edge to screen, respectively. Then Eq. 4.6 can be written

$$I = \left(\int \cos \frac{\pi(a + b)s^2}{ab\lambda}\, ds \right)^2 + \left(\int \sin \frac{\pi(a + b)s^2}{ab\lambda}\, ds \right)^2. \tag{4.8}$$

Let

$$\frac{\pi(a + b)s^2}{ab\lambda} = \frac{\pi}{2}v^2; \tag{4.9}$$

hence,

$$s = v\sqrt{\frac{ab\lambda}{2(a + b)}}, \tag{4.10}$$

which yields a total intensity

$$I_{total} = \frac{ab\lambda}{2(a + b)}\left[\left(\int \cos\frac{\pi}{2}v^2dv\right)^2 + \left(\int \sin\frac{\pi}{2}v^2dv\right)^2\right]. \tag{4.11}$$

The two integrals in the expression for the total intensity are the *Fresnel integrals*. The intensity at a point is, therefore, the sum of two integrals times a geometric term. Let

$$\sigma = \int \cos\frac{\pi v^2}{2}dv \tag{4.12}$$

and

$$\mu = \int \sin\frac{\pi v^2}{2}dv \tag{4.13}$$

These integrals cannot be evaluated in closed form except over the range 0 to ∞. They can, however, be evaluated by numerical methods. In the programs to follow Simpson's rule is used.

If σ and μ are evaluated and plotted for various values of v ranging from zero to infinity, the points lie on a spiral—the *Cornu spiral*. Each point on the spiral (σ, μ) corresponds to a value of v starting with $v = 0$ at the origin, and as v increases the point (σ, μ) moves along the spiral and equal increments in v advance the point equal distances along the spiral. Since the sign of v does not change the integral, only its sign, the spiral is symmetrical about the origin (Fig. 4.4). A mathematical study of the spiral forms one part of the programs. For an edge and the region inside the geometrical shadow (Fig. 4.5), a point on the direct line between the source O and a point on the screen P_2 is cut off by the obstruction. If s represents the arc along the wavefront from the point M_2, the pole of the wavefront, to the edge at A, then the intensity at P_2 is given by the Fresnel integrals (Eq. 4.11), with the limits s to ∞, or, in terms of v, from v to ∞ where

$$v = s\sqrt{\frac{2(a + b)}{ab\lambda}}. \tag{4.14}$$

Now if the distance from P_1, the point on the screen in line with the source O and the edge A, to P_2 is y:

$$y = \frac{a + b}{a}s \tag{4.15}$$

$$y = v\sqrt{\frac{(a + b)b\lambda}{2a}}. \tag{4.16}$$

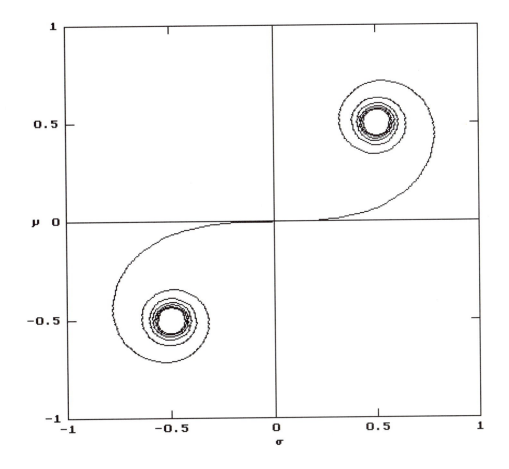

Figure 4.4: The Cornu spiral.

The Fresnel integrals integrated over the range zero to infinity can be evaluated directly to yield the value $\frac{1}{2}$. If

$$\int_0^v \cos \frac{\pi v^2}{2} dv = C_v \tag{4.17}$$

and

$$\int_0^v \sin \frac{\pi v^2}{2} dv = S_v, \tag{4.18}$$

then for a point in the shadow,

$$I = \frac{ab\lambda}{2(a+b)}\left[\left(\frac{1}{2} - C_v\right)^2 + \left(\frac{1}{2} - S_v\right)^2\right]. \tag{4.19}$$

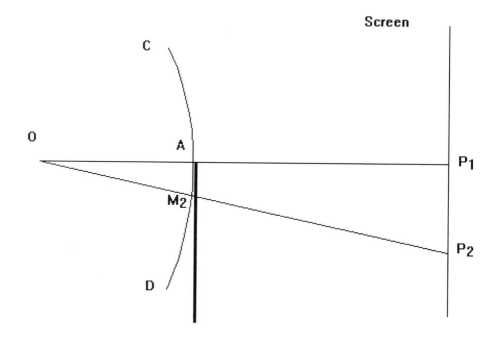

Figure 4.5: Diffraction at an edge: Inside the shadow.

For a point out of the shadow, above P,

$$I = \frac{ab\lambda}{2(a+b)}\left[\left(\frac{1}{2} + C_v\right)^2 + \left(\frac{1}{2} + S_v\right)^2\right]. \tag{4.20}$$

In relation to the spiral, I at a point y below P_1 is a function of the distance between the upper "eye," the point $\left(\frac{1}{2}, \frac{1}{2}\right)$, and the point on the spiral given by the appropriate value of v in Eq. 4.14 (e.g., point A [$v = 1.0$] in Fig. 4.6). As this point moves from the position where y is negative and large, i.e., well inside the shadow, toward the center, the point on the spiral moves along the curve, eventually reaching the origin $(0,0)$ at $v = 0$ (point B). As a result, the intensity increases gradually and smoothly from zero. As the point on the screen moves into the illuminated region the length of this line continues to increase until it reaches a maximum (point C [$v = -1.2$). Further increase in y causes the intensity to oscillate as the point winds around toward the lower eye. The amplitude of the oscillation decreases as this happens until a finite limit is reached, when the intensity is a function of the length of the line from $\left(\frac{1}{2}, \frac{1}{2}\right)$ to $\left(-\frac{1}{2}, -\frac{1}{2}\right)$.

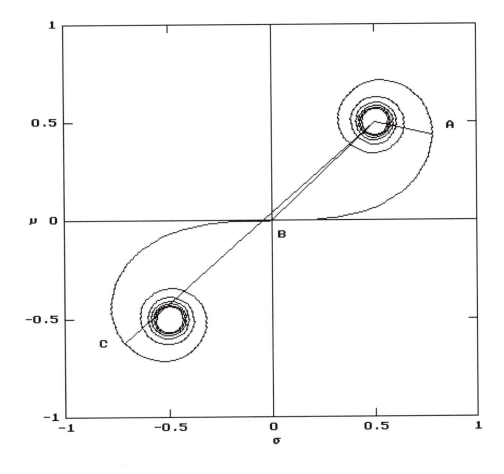

Figure 4.6: Diffraction at an edge.

In the case of a slit aperture (Fig. 4.7), the intensity is a function of a line joining two values of v, as defined by Eq. 4.14, these values being a function of the slit width and the position along the screen.

For a viewing point P_1 on the screen in a direct line from the source to the center of the slit, (Fig. 4.8), one point on the spiral is at the (σ, μ) value corresponding to the v value for y_{lower} and the other at the (σ, μ) corresponding to y_{upper} and $v_{upper} = -v_{lower}$ (e.g., points B [v = 0.8] and A [v = −0.8]). As the viewing point moves down, v_{lower} moves toward the upper eye and v_{upper} moves so that the increase in v_{lower} is matched by an equal decrease in v_{upper} and the arc length remains constant, such as points C [v = 1.0] and D [v = 2.6]. For a wide slit v_{lower} and v_{upper} would be well around the spiral towards the eyes and as a result the intensity pattern will show a series of oscillations across the screen. Eventually v_{upper} will itself move up toward the upper eye and the intensity will

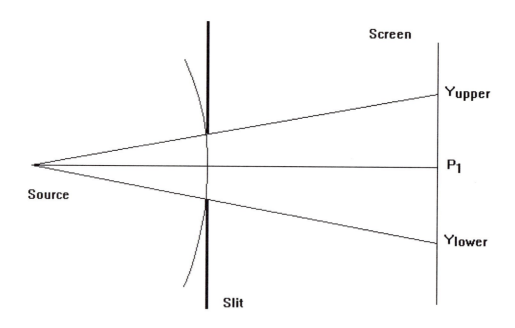

Figure 4.7: Geometry for diffraction by a slit.

fall smoothly to zero. As the point on the screen moves up, the opposite occurs and the eventually v_{lower} winds into the lower eye and the intensity pattern is symmetrical.

4.5 Fraunhofer Diffraction

4.5.1 Single Slit

The intensity pattern resulting from light falling on a slit can be examined by looking at the interference between the secondary wavelets sent out by every point along the wavefront as it reaches the aperture of the slit. The result[1] is given by

$$I = I_0(1 + \cos\theta)^2 \frac{\sin\beta^2}{\beta^2}, \qquad (4.21)$$

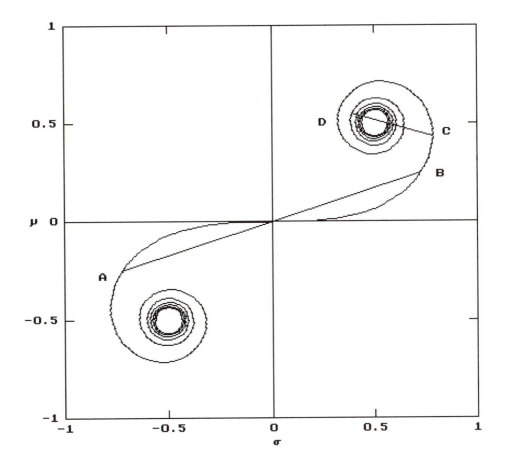

Figure 4.8: Diffraction by a slit.

where $\beta = \pi a \sin \theta / \lambda$, a is the slit width, θ is the angle of diffraction and λ is the wavelength of the incident light.

The intensity pattern shows a large central maximum I_0 and a series of smaller secondary maxima and minima across the screen. The minima are equally spaced and occur at values where

$$\beta = \pm k \text{ where k is an integer.} \tag{4.22}$$

The position of the secondary maxima are given by where $\tan \beta = \beta$.

As a result the maxima do not lie midway between the minima. Table 4.1[1] gives the values of k in the general expression

$$\sin \theta = \frac{k \lambda}{a}. \tag{4.23}$$

Table 4.1   Values of the constant k determining positions of maxima and minima for a single slit

Single-maxima and minima	
Maxima	Minima
0	1.000
1.430	2.000
2.459	3.000
3.471	4.000
4.477	5.000

Eq. 4.21 includes the obliquity factor. In practice, since θ is usually small, this factor has only a small effect on the amplitudes at large angles which are already small and will be left out of any subsequent expressions.

4.5.2   Multiple Slits

For an array of single slits, the analysis of the intensity pattern involves both interference from the set of slits and also diffraction from a single slit.[1] As a result the intensity function for this case is

$$I_\theta = I_0 \frac{\sin^2 \beta}{\beta^2} \cdot \frac{\sin^2 N\gamma}{\gamma^2}, \tag{4.24}$$

where $\gamma = \pi d \sin \theta / \lambda$, d is the slit separation, θ is the angle of diffraction, λ is the wavelength of the incident light, and N is the number of slits. The term involving β relates to the diffraction due to a single slit (Eq. 4.21); and the term in γ, the interference from the array of slits. The effect of adding slits is to enhance and sharpen some maxima and reduce the relative amplitude of others. There is a strong central maximum and a series of secondary maxima. The amplitudes of the secondary maxima vary with angle and every nth maxima is large and governed by the diffraction from a single slit. The other secondary maxima have much reduced intensities, the magnitude of the reduction being a function of N, the number of slits.

4.5.3   Rectangular Aperture

The problem of diffraction from a rectangular aperture, width w and height h, with a point source involves adding the effects of the spherical wavelets coming from a rectangular section of the wavefront defined by the aperture.[7] For such an aperture the intensity I is given by the expression

$$I = I_0 w^2 h^2 \frac{\sin^2 \beta}{\beta^2} \cdot \frac{\sin^2 \gamma}{\gamma^2}, \tag{4.25}$$

where I_0 is the incident intensity at the central maximum, λ is the wavelength, $\beta = \pi w \sin \theta / \lambda$, and $\gamma = \pi h \sin \phi / \lambda$,

The angles θ and ϕ are measured from the normal to the plane of the aperture at its center to the extremes of the width and height of the aperture. The intensity that results shows the typical one-slit variation of intensity along two perpendicular axes, the separation of the maxima being inversely proportional to the dimension of the aperture along that axis. The pattern also extends out into the four quadrants defined by the two axes but with much reduced intensity.

4.5.4 Circular Aperture

For a circular aperture, radius r, illuminated by a plane wavefront, the diffraction pattern displays a strong central maximum known as *Airy's disk*, surrounded by a series of much fainter rings. For such an aperture the intensity is given by the expression[5]

$$I_\theta = I_0 \left[\frac{2J_1\left(\frac{2\pi r \sin \theta}{\lambda}\right)}{\left(\frac{2\pi r \sin \theta}{\lambda}\right)} \right]^2, \tag{4.26}$$

where I_0 is the intensity at the central maximum, θ is the angle subtended by a point in the diffraction pattern by the center of the aperture, λ is the wavelength, and J_1 is the Bessel function of the first kind. The position of the minima amd maxima are functions of the Bessel function and Table 4.2[1] gives the values of m in the equation

$$\sin \theta_m = \frac{m\lambda}{r}. \tag{4.27}$$

4.6 *Coherence*

For interference effects to be created it is usually necessary for the interfering beams to be derived from the same source. In the case of two slits, a single source radiates waves that fall upon the slits and these then act as a pair of secondary sources. The important point here is there must be temporal stability of phase, or degree of phase correlation between the two wavetrains. The interfering wavetrains must have a phase relation which persists over the period of observation. Two independent sources will not usually show interference effects, because light coming from a source is not in the form of an infinite wave train. Such radiation comes in short bursts, which can be mathematically represented by a harmonic wave with a given frequency and wavelength. The wavetrains are finite in length and the phase relationships in the wavetrain suddenly change in a random fashion.

Table 4.2 Values of the constant m determining positions of maxima and minima for a circular aperture

Circular aperture—maxima and minima	
Maxima	Minima
0	1.220
1.638	2.233
2.666	3.328
3.674	4.241
4.722	5.243

This can, in fact, affect interference from two slits and a single source if the point on the screen being examined is in a position that the wavetrain reaching that point from one slit comes from a different wave group that generated the wavetrain coming from the other. This would mean the fringe pattern would disappear at points away from the central maximum. The time of passage associated with one of these wavetrains is the *coherence time*. It determines the region over which interference effects can be found.

Two basic schemes have been adopted to produce interference effects—division of wavefront, as in the two slit experiment, and division of amplitude, as in the Michelson interferometer discussed in chapter 6.

Lack of coherence can eliminate or reduce interference effects. The visibility of fringes in interference can also be affected in other ways. For example, if a single slit is illuminated by two *incoherent* line sources, the intensity pattern on a screen is the sum of the *intensities* on the screen There is no interference between the light coming from the two sources. The result depends on the separation between the two sources, and if they are close together the two central maxima can overlap to such an extent that there only appears to be one maximum. This effect can be very easily be seen if two laser beams illuminate the slit. No focusing lenses are then required. A similar effect can be seen with two point sources and a circular aperture and has a bearing on the ability of an instrument to separate two sources close together.

4.6.1   Single Slit—Two Line Sources

In the case of a single slit, the conditions for loss of visibility[2] depend on the separation distance between the sources, the distance that the sources are from the slits, the width of each slit, the separation between the slits, and the distance that the slits are from the screen. Each source will create an intensity pattern on the screen with its central maximum on the line joining the source to the slit. Because the sources are separated, the intensity maxima will be separated. The intensity patterns will add directly with no interference because the sources were taken to be incoherent. Whether it will be possible to see two maxima rather than one broad maximum or perhaps an intensity pattern with a saddle point between the maxima will depend on the separation between the sources and the slit width. It will be more difficult to see a distinction with a narrow slit because the widths of the intensity maxima are greater than with a wider slit. If the two maxima are visible they are said to be *resolved*.

An arbitrary criterion for the limit of resolution was defined by Rayleigh when the maximum of one pattern lies at the position of the first minimum of the other. For a single source and a single slit the angle subtended by the first minimum is given by (see Eq. 4.23)

$$\sin \theta = \frac{\lambda}{w} \tag{4.28}$$

corresponding to $\beta = \pi$, where w is the width of the slit. Hence, if the angle subtended by the two sources at the slit equals the angle as defined by Eq. 4.28 the diffraction patterns are said to be resolved. Under these conditions there is a small saddle point between the maxima. The two intensity curves cross at $\beta = \pi/2$ and the intensity due to one pattern is given by the relation

$$\frac{sin^2\beta}{\beta^2} = 0.405. \tag{4.29}$$

Hence the ratio of the intensity at the crossing point, the saddle point, to the maximum intensity will be 0.81. If the angle of separation between the two sources is less than the value of θ in Eq. 4.28, the interference patterns are said to be unresolved.

4.6.2 Circular Aperture—Two Point Sources

In a similar fashion to that discussed in the previous section, the limit of resolution of two point sources and a single circular aperture is given by the relation

$$\sin\theta = \frac{1.22\lambda}{R}, \tag{4.30}$$

where $R =$ the diameter of the aperture. The factor 1.22 relates to the position of the first maximum of the Bessel function. If the angle subtended by the two sources at the aperture is no less than this value, the two diffraction patterns are resolved on the Rayleigh criterion. It can, therefore, be seen that two sources would need to be a little further separated with a circular aperture than with a slit whose width was the diameter of the aperture.

4.6.3 Two Slits—Two Narrow Line Sources

In another example, consider two very narrow line sources illuminating two slits. Each source will create the familiar two slit diffraction pattern with one pattern shifted with respect to the other. For two very narrow slits the diffraction envelope due to each single slit will be very broad. The appearance of the fringe pattern will depend on the relative positions of the interference maxima and minima due to interference from the two slits. Since these fringes are equally spaced along the screen, the maxima created by one source may coincide with the maxima from the other, in which case clear fringes will be visible. If the maxima from one source lie at the minima from the other, no fringes will be seen. For other positions of the sources the fringes will have varying visibility.

Consider the diagram in Figure 4.9. Two sources, S_1 and S_2, are distance a from the two slits, SL_1 and SL_2. The slits are distance b from a screen. Each source and the two slits will create their own interference patterns on the screen. If P is a general point on the screen, the intensity at P which subtends an angle θ' with the x-axis will be given by the sum of two terms like

$$I = I_1[1 + \cos kd(\theta' - \theta_1)], \tag{4.31}$$

where I_1 is the intensity due to a single slit, θ_1 is the angle between the x-axis and the line joining source S_1 to the center of the slits, d is the distance of separation of the slits, and θ_1 will be defined as positive as it is above the horizontal axis.

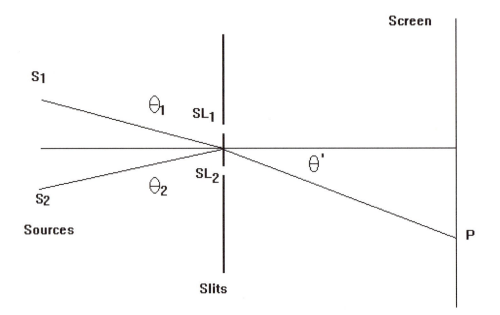

Figure 4.9: Two slits and two narrow line sources.

A similar expression will apply to the intensity on the screen due to source S_2, i.e.,

$$I = I_1[1 + \cos kd(\theta' - \theta_2)], \tag{4.32}$$

where θ_2, the angle between the x-axis and the line joining source S_2 to the center of the slits, will be negative. The total intensity will be the sum of these two terms. The maxima of one pattern will fall on the minima of the other when

$$kd(\theta_1 - \theta_2) = \pi \times \text{ odd integer} \tag{4.33}$$

or

$$\frac{d}{\lambda}(\theta_1 - \theta_2) = \text{ half integer,} \tag{4.34}$$

and the fringes will have minimum visibility. When

$$kd(\theta_1 - \theta_2) = 2\pi \times \text{ integer} \tag{4.35}$$

or

$$\frac{d}{\lambda}(\theta_1 - \theta_2) = \text{ integer,} \tag{4.36}$$

the fringes will have maximum visibility.

The *visibility function V* for the combination is defined as

$$V = \frac{I_{max} - I_{min}}{I_{max} + I_{min}},$$ (4.37)

and it can vary from zero to 1. Because the intensity pattern from two slits is a series of $\cos^2 \theta$ fringes which repeat across the screen, the condition in Eq. 4.34 will be met for many source separations. The first time that the fringes will disappear would be that where

$$\frac{d}{\lambda}(\theta_1 - \theta_2) = \frac{1}{2}.$$ (4.38)

A similar effect is seen in the Michelson interferometer, where this concept is exploited in measuring the separation between two spectral lines.

This phenomenon was also exploited by Michelson in his stellar inteferometer,[4] where he was able to measure the distance between two stars forming a binary star. He covered the objective lens of a telescope with a mask into which he cut two slits. As discussed above, the first time that the fringes disappear would be when

$$\theta_1 - \theta_2 = \frac{\lambda}{2d}.$$ (4.39)

For the known binary stars, the value of d that was required was greater than the diameter of the objective lens on the telescope. As a result Michelson mounted two mirrors at the opposite ends of a girder placed in front of the telescope which reflected light toward two more mirrors which reflected the light toward the slits. This had the effect of increasing the effective slit distance from d, the actual slit separation, to a larger value, the distance between the mirrors on the girder. Because of the very small intensities involved in this experiment, Michelson used white light from the binary star. The overlap of the fringes due to the non-monochromatic light meant that he was unable to see many fringes away from the central area, but he could see sufficiently well to allow the measurement to be made.

4.6.4   Two Slits—One Line Source

A third example of fringe visibility can be found if, in the two slit interference experiment, the two narrow line sources are replaced by one wide line source. This case is more complex in that the resultant fringe pattern is the combination of sets of fringe patterns coming from different parts across the line source.[2] Instead of simply adding two individual fringe patterns from two separate sources, it is necessary to integrate across the line source.

If we define $I(\theta)d\theta$ as twice the intensity due to a single slit coming from that part of the source between θ and $\theta + d\theta$, then

$$I_{\text{total}} = \int_{-\infty}^{\infty} I(\theta))d\theta.$$ (4.40)

For a line source of angular width $\Delta\theta$,

$$I(\theta) = I_{total}/\Delta\theta$$

$$I = I_{total}\left[1 + \frac{2}{kd\Delta\theta}\sin\left(\frac{kd\Delta\theta}{2}\right)\cos\left(kd(\theta' - \theta_0)\right)\right], \qquad (4.41)$$

where d is the slit separation and $\theta_0 = (\theta_1 + \theta_2)/2$.
     Hence,

$$I_{max} = I_{total}\left[1 + \mid \frac{\lambda}{\pi d\Delta\theta}\sin\left(\frac{\pi d\Delta\theta}{\lambda}\right)\mid\right], \qquad (4.42)$$

and

$$I_{min} = I_{total}\left[1 - \mid \frac{\lambda}{\pi d\Delta\theta}\sin\left(\frac{\pi d\Delta\theta}{\lambda}\right)\mid\right]. \qquad (4.43)$$

The visibility will then be

$$V = \mid \frac{\lambda}{\pi d\Delta\theta}\sin\left(\frac{\pi d\Delta\theta}{\lambda}\right)\mid . \qquad (4.44)$$

This function exhibits a series of maxima and minima as the width of the source increases, each successive maximum being smaller than the preceding one until the fringe visibility is very poor with a very wide source. The condition for a minimum value is given by

$$\frac{d\Delta\theta}{\lambda} = \text{integer}, \qquad (4.45)$$

which is just twice the value obtained for a double point source and a slit (Eq. 4.34).
     Michelson exploited this result in order to measure the diameter of stars. A modification is required if the star is assumed to be a circular disc and the conditions for the maxima and minima have to take the factor 1.22 into account.

4.7   Running DIFFRACT

4.7.1   General Comments

On startup you will see an introductory information screen. Pressing **Enter** or clicking the mouse will clear this screen and display the **Huyghens—Slits** program as shown in Figure 4.10. You can continue with this or select an alternative program by choosing from the top **Menu**.
     The first menu column **File** consists of five options. **About CUPS** provides general information on the CUPS project. The second option, **About Program**,

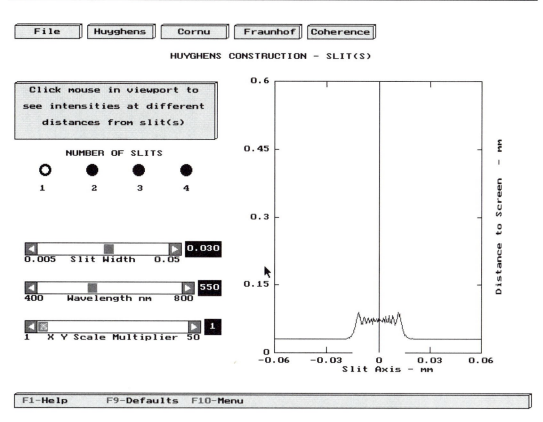

Figure 4.10: Huyghens: Slits.

provides a general introduction and credits. The third option, **Program Details**, provides some general information about the programs available to you. The fourth option, **Configuration**, controls system parameters and is fully discussed in chapter 1. The final option, **Exit Program**, allows you to leave the program. The various sections of the programs can be accessed using the other **Menu** options.

On the bottom of the screen is a set of hot keys. The **F1-Help** and **F10-Menu** hot keys are always available. Each program has its own set of hot keys for special purposes as appropriate. **F1-Help** provides context-sensitive help appropriate to the particular program you have chosen. **F10-Menu** allows you to activate the top line **Menu**. In addition you will find an **F9-Defaults** hot key, which runs the program with the original default values. Sometimes values are changed so much that it is often quicker to reset the defaults and then start again. Some programs have other hot keys which are discussed in the relevant section. On selecting a **Menu** item you will see that program for a set of default parameters. New parameters may then be set using the sliders/buttons. In the majority of cases the simulation runs on release of the slider thumb. A greater insight into many of the simulations can often be obtained by pressing the mouse button while the mouse arrow is located on either end of the slider. In this case the variable is smoothly changed and the plots change accordingly.

Menu Item: Huyghens

Slits

> **Default Settings:**
> Number of Slits—1
> Slit Separation—0.01 mm
> (for more than one slit)
> Slit Width—0.03 mm
> Wavelength—550 nm
> X-Y Scale Multiplier—1

If this **Menu** option is chosen, the opening screen shows the intensity on a viewing screen when a plane wave falls on one slit, as shown in Figure 4.10. The default plot appears after a short calculation time. The variation in the intensity function with distance from the slit(s) can be examined by clicking the mouse at any point within the viewport. The slit width and wavelength can be changed using the sliders. The number of slits can be selected using the buttons. If the number of slits chosen is greater than one, an extra slider is displayed that allows you to change the slit separation. Note that the horizontal and vertical ranges can be changed using the X-Y Scale Multiplier slider. This will allow you to see the intensity pattern at large distances from the slit(s).

In the calculation, the aperture of the slit(s) is filled with secondary radiating sources, one-eighth of a wavelength apart and, in this case, the obliquity factor is included (Eq. 4.2).

Note the way in which the shape of the intensity pattern across the screen varies as the screen is moved back from the slit(s). Close to the screen the intensity shows what is essentially a bright image of the slit(s). As the screen distance increases, however, the pattern changes slowly and a more familiar intensity pattern develops. For example, the Young's two slit experiment.

4.7.2 Menu Item: Huyghens

Point Sources

> **Default Settings:**
> Number of Point Sources—2
> Point Separation—550 nm
> ΔPhase—0°
> Wavelength – 550 nm
> X–Y Scale Multiplier—1
> Threshold—0.2

On selecting this **Menu** item the screen shows the intensity in the plane of the source(s). Sliders are provided so that you can change the wavelength, the phase difference between the radiating sources, the scale multiplier and the threshold value. A set of source buttons allows you to choose the number of sources. If the number of sources chosen is greater than one, an extra slider is displayed that allows you to change the source separation. In the calculation the intensity is calculated at each point in the plane of the sources using Eq. 4.4. The resulting matrix of intensity values is then displayed as a contour plot (11 uniform contour heights) and the radiating point(s) are marked with asterisks. The X-Y scale multiplier slider adjusts the range in the x and y directions simultaneously. The threshold slider allows you to increase the range of the calculation. It cuts off the intensity function at a predetermined height and then recalculates the contours over the reduced range. Any number of sources from one to eight can be selected. In the default setting all the sources radiate in phase – $\Delta phase = 0°$. The $\Delta phase$ slider allows you to adjust the relative phase between the sources over the range $-180°$ to $180°$.

The intensity patterns vary greatly as a function of the variables. Some of these features will be explored through the exercises that follow. However, you will notice that a single point gives the expected spherical distribution. When there is more than a single point the intensity pattern exhibits lobes, the number of lobes increasing with the number of sources. The position and strength of these lobes is then a function of the phase relationship between the sources.

4.7.3 Menu Item: Cornu

Study of the Spiral

The mathematical basis for the study of diffraction from a cylindrical wavefront was outlined earlier (page 104). In this program you can study some features of that analysis. On selecting this **Menu** item you will see, in the upper right viewport, the first section of the spiral. The two axes are σ and μ and the spiral is plotted up to $v = \pm1$. By using the **F2-Step** key you can advance the calculation by one increment in v up to $v = \pm5$. At any stage you can see the two curves that are integrated to calculate the σ and μ values for each value of v. To do this use the **F3-Cos** and **F4-Sin** hot keys. In each case, the range of the integration is indicated by the red line. If you study these curves carefully you will understand how the spiral develops. Also note that the distances measured along the spiral between integral values of v differing by unity are constant. The v values are marked with a red asterisk. By using the mouse in the viewport, you can measure σ and μ values for the integer values of v.

Edge

> **Default Settings:**
> Source to Edge Distance—0.1 m
> Edge to Screen Distance—1 m
> Wavelength—550 nm

In this program the intensity pattern across a screen is shown for a single edge. The position of the geometrical shadow is shown. The wavelength can be varied, as can the source to edge and edge to screen distances, but you will see that these two variables only affect the scale of the intensity pattern. In the calculation the intensity is a function of the distance from the upper eye to points on the spiral. As the position on the screen moves from well inside the shadow to the brightly illuminated part of the screen, the lower point winds along the spiral, reaching a maximum after a gradual and smooth increase. Once the first maximum is reached, the lower point winds around and causes the intensity to oscillate about a limiting value determined by the distance from the upper eye to the lower eye.

Single Slit

Default Settings:
Source to Slit Distance—0.1 m
Slit to Screen Distance—1 m
Slit Width—1.0 mm
Wavelength—550 nm

On selecting this **Menu** item, the calculation proceeds. Nothing appears on the screen until the calculation of the default case is complete. A **Calculating— please wait** reminder will display during this time. A typical screen is shown in Figure 4.11. This example is somewhat more complicated than the case of the single edge. The intensity is a function of the length of a line drawn from one point on the spiral (v value) which is a function of the position of one edge of the slit to another point on the spiral (a second v value) which is a function of the position of the other edge. As the point on the screen moves from one extreme to the other, this line adjusts according to the appropriate v values. The distance along the spiral between the ends of the line remains constant.

Sliders allow you to vary the slit width and the wavelength and the diffraction pattern is automatically recalculated and displayed. You can also move the source close to the slit, where the curvature of the wavefront will be greater, or far from the slit, where the incident wavefront approaches that of a plane wave. Similarly

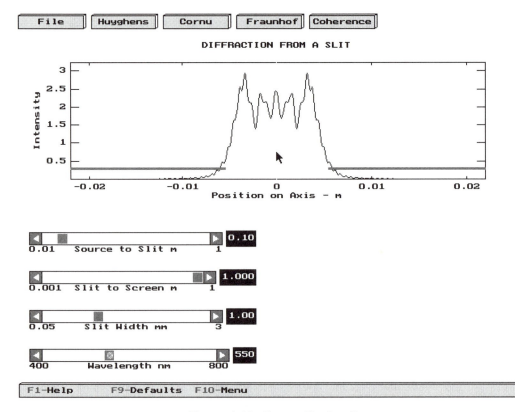

Figure 4.11: Cornu: Single slit.

you can move the screen close to the slit or far away. The spiral is calculated over a predetermined range of the v parameter. In this case the maximum value is 15. Certain combinations of the variable parameters may be selected that would require v values greater than this maximum value. In that event a blue message box is displayed asking for a change in parameter values. The message box indicates which way the parameters should be changed. If a slider is changed and the new maximum v value is appropriate, the calculation proceeds automatically.

By carefully adjusting the distances from source to slit with a narrow slit, you will be able to see the transition from Fresnel diffraction (source near and screen near) to Fraunhofer diffraction (source far and screen far). The position of the geometrical shadow is shown.

Single Obstacle

Default Settings:
Source to Obstacle Distance—0.1 m
Obstacle to Screen Distance—1 m
Obstacle Width—1.0 mm
Wavelength—550 nm

On selecting this **Menu** item there is a similar delay while the default calculation is made. This example is again somewhat more complicated than the case of a single edge. The intensity is a function of the length of two lines drawn on the spiral. One line joins the upper eye to a one point (one v value) which is a function of one edge of the slit. The other line joins the lower eye to another point on the spiral (a second v value) which is a function of the other edge of the slit. As the point on the screen moves from one extreme to the other, these lines adjust according to the v values. The distance along the spiral between the ends of the lines remote from the eyes remains constant. Thus one section of the spiral is always being cut out and the intensity is a function of the remaining sections of the spiral.

Control of the same variables as with the single slit is available to you. A similar limitation applies in terms of allowable parameter values as discussed in the case for a single slit. The position of the geometrical shadow is shown. Again you can study the gradual transition from Fresnel diffraction to Fraunhofer diffraction. If you use a narrow obstacle you will be able to see that there is a bright spot in the centre of the geometrical shadow, similar to the Poisson spot obtained with a circular obstacle.

Double Slit

Default Settings:
Source to Slits Distance—0.1 m
Slits to Screen Distance—1 m
Wavelength—550 nm
Left Slit Width—1.0 mm
Separation Factor—1 (1.0 mm)
Right Slit Factor—1 (1.0 mm)

The calculation time for this program is somewhat longer than for the previous cases. This is a function of the more complex analysis required and a larger v range. In this case the spiral is divided into sections. The first part involves a line from a point on the spiral which is a function of the position of the left edge of the left slit to another point which is a function of the position of the right-hand edge of the left-hand slit. A second line runs from a point on the spiral which is a function of the position of the left-hand edge of the right-hand slit to a point that is a function of the position of the right-hand edge of the right-hand slit. The intensity at a point on the screen is then a function of these two lines drawn on the spiral taking into account their magnitudes and directions.

You can choose freely, over a given range, the width of the left-hand slit. As a contraint of the method of calculation, however, the slit separation and the width of the right-hand slit have to be an integer function of the left slit width. These variables are controlled by the Separation Factor and right slit width integer sliders. This limitation is not serious.

Again, there are limations on appropriate parameter values. The positions of the geometrical shadows are shown. The same comments made above also apply here in terms of the transition from Fresnel to Fraunhofer diffraction. Can you set up Young's two slits?

Double Obstacle

> **Default Settings:**
> Source to Obstacles Distance—0.1 m
> Obstacles to Screen Distance—1 m
> Wavelength—550 nm
> Left Obstacle Width—1.0 mm
> Separation Factor—1 (1.0 mm)
> Right Obstacle Factor—1 (1.0 mm)

After the required calculation time, the intensity due to two obstacles is shown in the upper viewport. The analysis follows the calculation for a single obstacle except that the intensity is now a function of three sections on the spiral suitably compounded. The first line runs from the upper eye to a point on the spiral which is a function of the position of the left edge of the left obstacle. Another line runs from another point on the spiral which is a function of the position of the right-hand edge of the left-hand obstacle and the position of the left-hand edge of the right-hand obstacle. Finally, a third line runs from a point that is a function of the position of the right-hand edge of the right-hand obstacle to the lower eye. The intensity at a point on the screen is then a function of all three lines drawn on the spiral, taking into account their magnitudes and directions.

Again there are limitations on appropriate parameter values. The positions of the geometrical shadows are shown.

4.7.4   Menu Item: Fraunhofer

Single and Multiple Slits

> **Default Settings:**
> Number of Slits—2
> Slit Width—0.02 mm
> Slit Separation—0.2 mm
> Wavelength—550 nm
> Angular Range—2.5°

On selecting this **Menu** item the intensity for the default case is plotted as a function of angle. The slit width and the wavelength can be changed using the sliders. The number of slits can be selected using the buttons. If more than one slit is chosen, a slider is displayed that can be used to change the slit separation. You can display more than one intensity plot at a time by selecting new parameter values which is useful in making comparisons. A special hot key, **F3-Clear All**, allows you to clean the viewport, and the **F2-LastPlot** hot key recalculates and displays the simulation for the most recent slider and buttons settings. The angular range is also controlled by a slider. If the angular range is changed, the viewport is cleaned and the last graph is replotted immediately on release of the slider, using the new angular range. If the mouse is moved into the viewport, the intensity can be measured as a function of position. The calculations are based on Eqs. 4.21 and 4.24. You will be able to make comparisons with the results that you saw in the programs dealing with Huyghens construction and the Cornu spiral.

Rectangular Aperture

> **Default Settings:**
> Threshold—0.1
> Width—0.02 mm
> Height—0.02 mm
> Wavelength—550 nm
> Screen Distance—1 m (fixed)

On selecting this **Menu** item the opening screen shows, in the lower right-hand viewport, a three-dimensional intensity plot. The same information is shown in the lower left-hand viewport as a contour plot. Sliders are available to change the parameters. One slider controls the intensity function. The central maximum is very intense compared to the secondary maxima and as a result these secondary maxima are difficult to see. The **threshold** slider allows you to cut off the main central maximum at any desired level consequently magnifying the secondary features. Another facility is available that will allow you to look at the intensity over a different field of view. A dummy screen sits in the upper left. By clicking the mouse inside the red square and then dragging the square to a new position, the calculation is repeated for a new screen of the same size but centered at the new position. The new coordinates of the center of the field of view are displayed in the blue message box and the calculation begins when the mouse button is released. The calculations are based on Eq. 4.25.

You should also study the Fourier transform programs in chapter 2. In particular look at the 2D Fourier transform of a rectangular aperture.

Circular Aperture

> **Default Settings:**
> Threshold—0.1
> Radius—0.01 mm
> Wavelength—550 nm
> Screen Distance—1 m (fixed)

This option is similar to the rectangular aperture above. The screen layout is the same and a similar set of sliders controls the parameters. The intensity pattern is a set of rings around a strong central maximum. The calculations are based on Eq. 4.26.

As was suggested above, you should also study the Fourier Transform programs in chapter 2. In particular look at the two-dimensional Fourier transform of a circular aperture.

4.7.5   Menu Item: Coherence

Circular Aperture—Two Point Sources

Default Settings:
Source Separation—10 mm
Radius—0.04 mm
Wavelength—550 nm
Sources to Aperture Distance—1 m (fixed)
Aperture to Screen Distance—1 m (fixed)

In the lower left viewport of the opening screen the individual intensity patterns for each source and a circular aperture are displayed as a function of angle. Also shown is the sum of the two intensities. A calculation is also made to display the information as a three-dimensional plot which is shown in the lower right-hand viewport. This calculation takes place while the **Calculating—please wait** message shows. Sliders can be used to change the variables and the 3-dimensional plot can also be rotated about two axes for closer examination of the features. To increase the range of use of this program, three buttons are placed on the upper right hand side of the screen. This allows you to examine the resolution of three different ranges of aperture—0–0.08 mm, 1–5 mm and 100–1000 mm. Choosing another aperture automatically changes the range of the source separations that can be examined.

In the exercises you will be able to study the resolution of the human eye and of telescopes. If the mouse is clicked inside the box in the left-hand corner of the three-dimensional plot, a contour plot is displayed. The original three-dimensional plot can be redisplayed by clicking a second time in the corner box. If the mouse is moved into the lower left-hand viewport, it is possible to measure the intensity as a function of position. If you measure a maximum and the minimum at the saddle point you will be able to calculate the ratio and check against the Rayleigh criterion. Displayed at the top of the left hand viewport is the angle of separation subtended by the sources at the aperture, in arcseconds.

Single Slit—Two Line Sources

Default Settings:
Source Separation—10 mm
Slit Width—0.08 mm
Wavelength—550 nm
Sources to Slits Distance—1 m (fixed)
Slits to Screen Distance—1 m (fixed)

The opening screen is similar to the case above for a circular aperture. Sliders allow you to change the variables and the mouse can again be used to measure the intensity as a function of position.

Two Slits—Two Narrow Line Sources

> ### Default Settings:
> Wavelength—550 nm
> Source Separation—0.01 m
> Slit Width—0.002 mm
> Slit Separation—0.03 mm
> Sources to Slits Distance—1 m (fixed)
> Slits to Screen Distance—1 m (fixed)

When this **Menu** option is chosen, the opening screen displays three viewports and a set of four sliders. The upper viewport shows the intensity pattern on the screen due to one of the sources and the two slits. The middle viewport shows a similar intensity pattern for the second source. The lower viewport shows the total intensity resulting from the addition of the two individual intensity patterns.

The two sliders on the left side allow changes to be made in the source parameters. The other pair of sliders on the right control the slit variables. Both separation sliders have a **Zoom** facility built in to allow for fine control of these parameters.

With the two sources located at the same point, maximum visibility fringes are observed. As the two sources are moved apart, the visibility falls until at the critical separation distance given by Eq. 4.34 it becomes zero. Further increase in the source separation creates fringes with increasing visibility until a maximum is reached according to Eq. 4.36. This cycle then repeats as the sources move further apart. Note that a finite coherence time would eventually cause the individual fringes making up the combined intensity to disappear. This program does not factor in this effect.

Two Slits—One Line Source

> ### Default Settings:
> Wavelength—550 nm
> Source Width—0.04 m
> Slit Width—0.002 mm
> Slit Separation—0.03 mm
> Sources to Slits Distance—1 m (fixed)
> Slits to Screen Distance—1 m (fixed)

On selecting this option from the **Menu**, the opening screen displays an upper viewport in which is plotted the visibility function V as a function of the parameter $d\Delta\theta/\lambda$. On the left-hand side of the screen are the source sliders which control the wavelength of the light and the width of the source. On the right-hand side are the slit sliders that control slit width and slit separation. The source width slider and the slit separation slider have a fine adjustment **Zoom Range** facility.

The default calculation, which involves integration across the source width, proceeds on acceptance of the **Menu** item and during this time the **Calculating— please wait** sign is displayed. The result is then displayed in the lower viewport. Also shown above this lower viewport is the value of the function $d\Delta\theta/\lambda$, which controls the visibility. By adjusting the source width and/or the slit separation and/or the wavelength it is possible to cause the fringes to disappear and this will occur whenever the function $d\Delta\theta/\lambda$ has an integral value. The value of the function $d\Delta\theta/\lambda$ is updated if the relevant sliders are changed, on release of the slider thumb.

The mouse can be used in the lower viewport to measure the intensity at points on the curve. If you measure at a maximum and then a minimum you will be able to note these values and calculate the visibility function using Eq. 4.37. In the upper viewport is plotted the variation of the visibility as a function of $d\Delta\theta/\lambda$. The green line is plotted at the current value of $d\Delta\theta/\lambda$ and the values of visibility and $d\Delta\theta/\lambda$ are provided.

Double Star

Default Settings:
Default Double Star—Castor
Wavelength—550 nm
Mirror Separation—0.025 m
Slit Width—0.002 mm (fixed)
Slits to Screen Distance—1 m (fixed)

This **Menu** option uses the ideas developed in the earlier program—Two Slits/Two Line Sources. On selecting this **Menu** option, the viewport shows the intensity across the field of view for the double star **Castor** for the default parameters. Fringes are visible. There are two sliders: on the left there is a slider to see the effect of changing the wavelength of the incident light and on the right one for adjusting the mirror separation.

The star **Castor** has been preselected using the buttons on the upper portion of the screen. Beside the star name is a value for the separation in astronomical units (AU). This value is only valid for fringes of zero visibility. Also listed is the angular separation as defined by the current parameter values. The value of the angular separation is updated on release of the slider thumb or during a change if you use the mouse cursor at the end of the slider. In order to measure the actual separation in AU and the angular separation in seconds of arc, it will be necessary to adjust the mirror separation from its smallest value in an attempt to reduce the visibility to zero. When this is achieved, and fine adjustment of the mirror separation is available using the **Zoom** facility, the calculated values are valid. Note that it may be possible to find larger mirror separation values for zero fringe visibility.

Three other double stars can be studied by choosing a new double star, using the buttons provided. These will form part of the exercise set.

Star Diameter

Default Settings:
Default Star—Betelgeuse
Wavelength—550 nm
Mirror Separation—1 m
Slit Width—0.002 mm (fixed)
Slits to Screen Distance—1 m (fixed)

This program is based on the Two Slits/One Line Source program. The only difference here is that, instead of a line source, the analysis assumes that the single star can be considered to be a circular disc. The factor 1.22 is, therefore, included.

The opening screen is somewhat similar to the double star application of the stellar interferometer. The default star is **Betelgeuse**.

In the lower viewport is the appearance of the fringes in the field of view for the default parameters. Beside the star name is the diameter in Sun diameters. Also shown is the value for the arc subtended by the star for the current values and the value of the function $d\Delta\theta/\lambda$. The values of these functions are updated on release of the slider thumb or during a change if you use the mouse cursor at the end of the slider. These two values are only correct if the fringe visibility is zero. This will you assist you in finding the settings for zero fringe visibility.

To measure the star diameter, choose a wavelength representative of the incident light and move the mirror from its smallest separation until the fringes vanish. The appropriate values of the star diameter and arc are displayed. You may find that it is possible to set more than one mirror separation that gives zero fringe visibility. If so you will notice that the visibility maxima between the minima falls according to the visibility graph. The visibility graph that you saw in a previous program is not appropriate unless you factor in 1.22 on the abscissa. Also note that larger mirror separation values will not lead to correct diameter values unless you make a correction for the order of the maximum.

The other stars can be studied using the buttons and this forms part of the exercise set to follow.

4.8   Exercises

Huyghens

4.1   Slits
a. Choose a single slit. Study the way in which the intensity varies with distance from the screen and as function of the width. Also study the effect of changing the wavelength.
b. Repeat this exercise for two, three, and four slits. Make sure that you explore a wide range of screen distances and screen widths. (Note that the calculation time is proportional to the number of slits.)

4.2   Point Sources
a. Study the manner in which the diffraction pattern varies with the number of points for a fixed phase (0°). Note the lobe structure that develops as you increase the number.
b. Set the number of points greater than one. Study the intensity variation as the phase difference is varied from 0° through to 360°.
c. Set the phase difference to 0° and study the shape of the lobe structure as the number of points is increased. (Note that the calculation time is directly proportional to the number of points.)
d. In a phased radar linear array,[6] the elements of the array behave as dipoles with their axes parallel. The radiation from each dipole shows spherical symmetry in a plane perpendicular to the dipole direction. By now you will probably have realized that, if you have looked at the previous questions, the lobes move in space with the phase difference.

This has been exploited in the the phased antenna array where by electronically controlling the phase difference, a beam of radiation can be made to sweep around through 360°. This saves having to rotate the whole array, which, if there are many elements, can be a formidable mechanical problem.

Use the program to study the properties of such an array. Of course, the wavelengths available to you are not representative of radar waves. However, the program scales directly and the results will be valid.

Study of the Spiral

4.3   Cornu

a. Study the way in which the spiral builds as you increase the v value. Measure the σ and μ values as a function of v.

b. Study the Fresnel integrals so that you can understand the way in which the spiral builds as function of v.

4.4   Edge

a. Study the effect of wavelength change on the diffraction pattern. From this information predict the effect of using white light as a source.

b. Why does the pattern not change with screen distance and source distance?

4.5   Single Slit

a. Study the manner in which the variation in the diffraction pattern for a single slit set at the default setting varies with the distance from the source to the slit and with the distance from the slit to the screen. Note the transition from Fresnel to Fraunhofer conditions.

b. How does the pattern vary with wavelength under both conditions?

c. Study the variation in the diffraction pattern for a range of slit widths under both Fresnel and Fraunhofer conditions. Relate these findings to the spiral so that you are clear as to why the transition occurs.

4.6   Single Obstacle

a. Study the variation in the diffraction pattern for a single obstacle set at the default setting varies with the distance from the source to the obstacle and with the distance from the obstacle to the screen. What do you believe the transition from Fresnel to Fraunhofer will be like? Are your ideas confirmed by your observations?

b. How does the pattern vary with wavelength under both conditions?

c. Study the variation in the diffraction pattern for a range of obstacle widths under both Fresnel and Fraunhofer conditions. Relate these findings to the spiral so that you are clear as to why the transition occurs.

4.7 Double Slit
   a. Using the default settings, study the changes in the diffraction pattern for two slits as you vary the source to the slits and the slits to screen distances. As in the case of a single slit, note the transition from Fresnel to Fraunhofer conditions.
   b. Make a similar study as a function of slit widths and separation. Set the slits at the same width and adjust the separation to confirm the previous results using the Huyghens program.
   c. Continue the study in the last question using different slit widths.
   d. As you answer these questions make sure that you can see how the results relate to the spiral.
   e. Study the effect of changing the wavelength. What would you expect if a white light source was used?

4.8 Double Obstacle
   a. Using the default settings, study the changes in the diffraction pattern for two obstacles as you vary the source to the obstacles and the obstacles to screen distances. As in the case of a single obstacle, note the transition from Fresnel to Fraunhofer conditions.
   b. Make a similar study as a function of obstacle widths and separations.
   c. As you answer these questions make sure that you can see how the results relate to the spiral.
   d. Study the effect of changing the wavelength. What would you expect if a white light source was used?

Fraunhofer

4.9 Single and Multiple Slits
   a. Study the diffraction pattern from a single slit and the interference pattern from more than one slit. Use the mouse to measure the intensity function as a function of angular position. Confirm that the intensity of the maxima are governed by Eq. 4.21.
   b. If the ratio of slit width to slit separation takes certain values some of the interference maxima are "missing." Study this effect for two slits and establish the relationship.
   c. Investigate the wavelength dependence of the intensity function.
   d. Study the effect of increasing the number of slits over the available range. Does the missing order relationship still hold?

4.10 Rectangular and Circular Apertures
   a. Study the diffraction pattern for a square aperture and a circular aperture over as a wide a range of field of view as you can.
   b. Investigate the effect of changing the ratio of the length of the sides of the rectangle and the radius of the circle.
   c. What is the effect of changing the wavelength?
   d. Study the Fourier transform for a rectangular and a circular aperture using the FOURIER program.

Coherence

4.11   Circular Aperture—Two Point Sources
   a. Study the way in which the combined intensity is a function of the separation distance, the size of the aperture, and the wavelength.
   b. Relate these observations to the Rayleigh criterion.
   c. A pair of point sources emit incoherent green light. They are placed 1 m from a circular aperture. If the wavelength of the light is 500 nm, what is the minimum radius that would allow you to be able to just resolve the two sources using the Rayleigh criterion? How far apart would the sources have to be to be resolved if they were red (700 nm)?
   d. Calculate the theoretical resolving power (in seconds of arc) for the human eye in bright light and dark conditions.
   e. Using the information from the previous question, what is the minimum separation of two bright point objects on the surface of the Earth that could be distinguished by a scientist in the *Shuttle* vehicle at a height of 500 km?
   f. The largest refracting telescope is at the Yerkes Observatory and it has a lens of diameter 1.016 meters. What is its theoretical resolving power?
   g. Design a telescope capable of seeing and resolving the two stars in the double star *Alpha Centauri*. (See the Michelson stellar interferometer questions that follow).

4.12   Single Slit—Two Line Sources
   a. Study the way in which the combined intensity is a function of the separation distance, the slit separation, and the wavelength.
   b. Relate these observations to the Rayleigh criterion
   c. The two sources in the third question in the preceding section were replaced by two line sources. Redo that question if the circular aperture was replaced by a single slit.

4.13   Two Slits—Two Narrow Line Sources
   a. Study the way in which the total intensity pattern varies as the distance between the two sources is increased and as the slit width is changed.
   b. Repeat this exercise for a different wavelength.
   c. Two slits, 0.03 mm apart and 0.0002 mm wide, are 1 m from two sources radiating light at 600 nm. Find as many separation distances as you can that cause the fringes to disappear. Show that these distances are in a simple ratio to each other. What results do you get with green light at 500 nm?
   d. Repeat the last question for a slit separation of 0.01 mm.

4.14   Two Slits—One Line Source
   a. Study the total intensity pattern as the width of the line source is changed and the wavelength is varied.
   b. A line source of width 0.02 mm emits red light of wavelength 650 nm and it falls on a double slit 1 meter away. The slit widths are 0.0002

mm. Find as many slit separation values that give fringes of maximum visibility. Make a series of measurements of the visibility function as a function of slit separation and confirm you results directly from the visibility graph.

    c. Repeat the last question for green light of wavelength 475 nm.

4.15   Double Star

    a. Measure the separation distance between the components of the double stars **Castor, Kruger 60, pEri,** and **Alpha Centauri**. If possible, find more than one setting of the mirror that gives zero fringe visibility and make the appropriate adjustment to the calculation of the separation distance.

    b. Repeat the exercise for another wavelength and make some estimate of the problem that would arise if monochromatic light was used (as was the case in the original experiment).

    c. What is the maximum separation distance that you could measure with the interferometer?

4.16   Star Diameter

    a. Measure the diameter of **Betelguese, Aldeberan,** and **Arcturus**. Find as many mirror settings as you can for each star.

    b. What is the maximum angular diameter that could be measured with the interferometer.

    c. What modifications, if any, would you need to make to measure the diameter of **Sirius**?

4.9   Further Developments Involving Optional Programming

If you are going to attempt the advanced programming exercises below, make sure that you have made copies of the original code. Keep these originals safe and only develop the copies. Do not be deterred if a new program does not run or it crashes. You might be wise to study the *CUPS Utilities Manual.*

    a. Develop code to allow for a study of coherence time and coherence length.

    b. Develop the Fraunhofer program to cover the case of multiple apertures. This could be a set of rectangular and/or circular apertures in a regular array and then in an irregular array.

References

1. Jenkins, F.A., and White, H.E. *Fundamentals of Optics.* New York: McGraw-Hill, 1976.

2. Klein, M.V. *Optics.* New York: John Wiley and Sons, 1970.

3. Lommel, E.V. *Bayerischen Akademie der Wissenschaften* **15**:531, 1886.

4. Michelson, A.A. On the application of interference methods to astronomical measurements, Astrophysical Journal **51**:257, 1920.

5. Preston, T. *Theory of Light.* London: MacMillan, 1928.

6. Shen, L.C., and Kong, J.A. *Applied Electromagnetism.* Monterey, CA: Brooks/Cole, 1983.

7. Wood, R.W. *Physical Optics.* London: MacMillan, 1905.

5

Ray Tracing in Geometrical Optics

Brian W. James

5.1 Introduction

There are two programs concerned with tracing rays through optical systems, RAY-TRACE and TWOLENS. The program RAYTRACE illustrates the propagation of light rays through various materials and lens systems, allowing the user to change parameters and observe the effects, thus enhancing the understanding of these topics.There are many different options in the program: namely, **Fermat, Mirage, Fiber, Rainbow,** and **Lenses**. All of these options, which are based on the geometrical approach and assume that light travels in straight rays, can be used whenever the geometrical dimensions involved are large compared with the wavelength of the light. The program TWOLENS illustrates simple ray tracing for thin lenses, two lens combinations, and reflection by spherical mirrors.

All the calculations for the propagation of light through various materials in the program RAYTRACE are based on Fermat's principle which states that the path taken by a ray is such that the time taken is at a stationary value. This stationary value is usually a minimum.[2,3] The *optical path* is defined as the product of refractive index n and geometrical distance s. The refractive index is defined as the ratio of the speed of light in free space (vacuum) to the speed of light in the medium.

For a ray passing through several different transparent materials, the total optical path, W, between two points A and B, is given by

$$W = \sum_{j=1}^{\infty} n_j s_j, \tag{5.1}$$

where the summation is taken over all the path segments, j, of length s_j and of refractive index n_j. In the case of a continuous variation of refractive index with position, the summation may be replaced by an integration with s_j replaced by the element of distance ds so that

$$W = \int_A^B n(s)\,ds, \tag{5.2}$$

where $n(s)$ indicates the dependence of the refractive index n on position. The total optical path is the equivalent distance the light would travel in a vacuum.

The basic laws of reflection and refraction can be deduced from Fermat's principle.

5.2   Refraction at a Plane Surface

Consider the path of a ray of light, shown in Figure 5.1, from A to B, which is refracted at D on the refracting surface RR'. The plane refracting surface RR' separates media of refractive indices n above RR' and n' below RR', with $n' > n$. The angle of incidence is i and the angle of refraction is i'.

It is convenient to let D be at the origin of Cartesian axes so that B is a distance x in the x direction from D. The point A is at $((c-x), a)$, B is at $(x, -b)$ and the separation of A from B in the x direction is c. The time, t, taken by the ray in travelling from A to B is given by

$$t = \frac{AD}{v} + \frac{DB}{v'} = \frac{\sqrt{a^2 + (c-x)^2}}{v} + \frac{\sqrt{b^2 + x^2}}{v'} \tag{5.3}$$

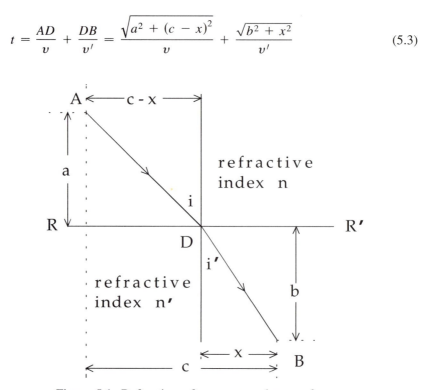

Figure 5.1: Refraction of a ray at a plane surface.

where v is the velocity of light in the material above RR' and v' is the velocity of light in the material below RR'. Thus, applying Fermat's principle, the first derivative of time with respect to path is

$$\frac{dt}{dx} = \frac{-(c - x)}{v\sqrt{a^2 + (c - x)^2}} + v^1\frac{x}{\sqrt{b^2 + x^2}} = 0. \tag{5.4}$$

Therefore

$$\frac{\sin i'}{v'} = \frac{\sin i}{v}, \tag{5.5}$$

so that

$$n \sin i = n' \sin i'. \tag{5.6}$$

This is Snell's law of refraction. For parallel layers, of transparent media of different refractive index, it can be written in the general form

$$n'_j \sin i'_j = k, \tag{5.7}$$

where k is a constant for a particular ray, and for layer j the refractive index is n'_j and i'_j is the angle of refraction. In the program RAYTRACE the option **Fermat** provides an opportunity to test the agreement of Fermat's principle and Snell's law and the **Mirage** option illustrates an application of Eq. 5.7.

5.3   Total Internal Reflection

The initial assumption in the previous section that $n' > n$ ensured that the incident ray always produced a refracted ray and that $i' < i$. As a consequence of the reversibility of rays of light we can have situations in which $n > n'$ and $i' > i$, where the undashed value refers to the incident ray and the dashed value refers to the refracted ray. In these circumstances it is possible that for some values of i

$$\frac{n \sin i}{n'} > 1, \tag{5.8}$$

to which there is no solution for i' from Eq. 5.6 since the argument of $\sin^{-1}$ cannot exceed unity. In this case there is no refracted ray and all the light is reflected. The angle of reflection is equal to the angle of incidence and the ray is said to suffer total internal reflection. In the simulation **Fermat** the conditions for total internal reflection can be examined. In the screen in Figure 5.2 the **Fermat** option display is shown. The user can adjust the sliders to investigate Fermat's principle. The components of the optical path in Eq. 5.1 are shown, and also the total optical path in separate graphs.

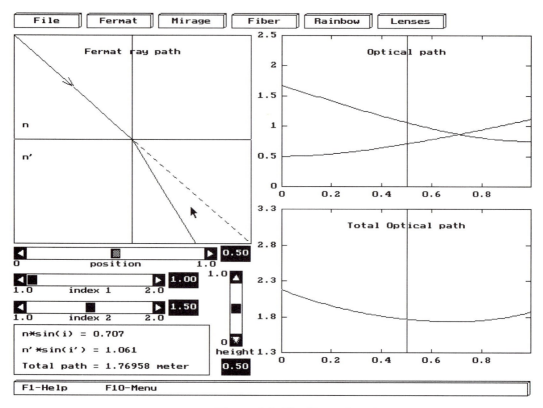

Figure 5.2: The **Fermat** option.

5.4 Formation of a Mirage

A common situation in which a mirage is observed is when a dry road appears to be wet due to the reflection of light from the sky near the road surface. This phenomenon is usually observed along straight, nearly flat roads on hot sunny days.[2] The air above the road becomes heated due to absorption of the sun's energy by the road surface, which heats the air adjacent to its surface; and if there is only a little wind, this hot air rises slowly above the road's surface. The normal small refractive index of air is reduced further when the air is less dense due to increased temperature at constant pressure. The lowest refractive index will occur at the road surface where the air is hottest. According to the law of Gladstone and Dale quoted by Longhurst[12] and Guenther[3],

$$(n - 1) \propto \rho, \tag{5.9}$$

where ρ is the density of the air. Under conditions of constant pressure, it can be seen from the gas law,

$$PV = nRT, \tag{5.10}$$

that

$$\rho = \frac{nR}{V} = \frac{P}{T}; \tag{5.11}$$

so that

$$(n - 1) \propto \frac{1}{T}. \tag{5.12}$$

The air above the road may be considered to be in layers, each layer with a different refractive index. As rays of light from the sky pass from one layer to the next, their direction of travel will be changed. The general form of Snell's law in Eq. 5.7 provides a convenient starting point to calculate the path of a ray. The formation of a mirage involves reflection at some layer in the air above the heated surface, which occurs when the condition in Eq. 5.8 is satisfied. In this case the critical angle is effectively $\pi/2$ since the layers of different refractive indices are infinitesimally thin, and the condition in Eq. 5.8 is satisfied when $i_j \rightarrow \pi/2$ as $n_j' \rightarrow n_j$. The paths of rays which give rise to a mirage are illustrated in the program RAYTRACE in the **Mirage** option. The screen in Figure 5.3 is taken from the program with the **Mirage** option selected. The user can adjust the slider to investigate the paths of rays at various angles below the horizontal which enter the eye of an observer.

5.5 Optical Fiber Design

Optical fibers used in telecommunications systems depend upon total internal reflections for their operation. Light entering at one end of a long thin fiber emerges

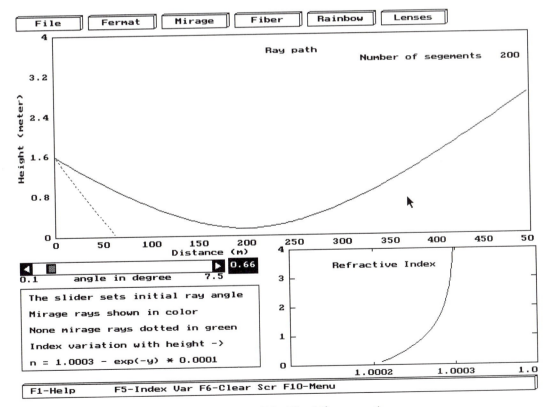

Figure 5.3: The **Mirage** option.

at the other end after undergoing multiple internal reflections along the fiber. The intensity of light passing through the fiber must not be reduced rapidly, so low-attenuation glasses have been developed. In a simple fiber, light will be lost at the support points and give an apparently higher attenuation. There are two methods adopted to overcome this effective attenuation loss, both of which involve making a composite fiber. One solution is the step-index fiber in which a central glass core is surrounded by an outer glass of a lower refractive index. Total internal reflection occurs at the boundary between the core and the outer glass so that the support of the outer glass does not affect the performance of the system.

The second solution adopted is to make a graded-index fiber in which the composition of the glass varies in a radial manner so that the center has a higher refractive index than the outer regions. In this type of fiber total internal reflection occurs in a manner similar to that in which it occurs in the mirage phenomenon. It is possible to gain some understanding of the design problems of step index fibers for telecommunications by using the ray tracing methods but with two additional concepts being required. A full Maxwell's equation electromagnetic wave treatment is given by Wilson and Hawkes.[4] The conditions for total internal reflection have been introduced in Eq. 5.8. It is necessary, in considering optical fiber design, to know more about the reflected light. There is some reflected light at all angles of incidence, so that Figure 5.1 should be modified by the addition of a reflected ray at angle i to the normal to the surface. The amplitude and phase of the reflected and transmitted light relative to the incident light depends upon the plane of polarization of the light as well as the angle of incidence. In particular at a boundary between a region with refractive index n and a region with refractive index n' for light with the plane of polarization perpendicular to the plane of incidence it has been shown[4] that the phase change 2ϕ on reflection is given by

$$\tan \phi = \frac{\sin^2 i - \left(\frac{n'}{n}\right)}{\cos i}.$$

(5.13)

For light with the plane of polarization parallel to the plane of incidence it has been shown[4] that the phase change 2δ on reflection is given by

$$\tan \delta = \left(\frac{n}{n'}\right)^2 \tan \phi.$$

(5.14)

In both cases the phase changes on reflection lead the phase of the incident light.

The simplest form of optical waveguide is a slab of glass of thickness d and refractive index n, sandwiched between two semi-infinite regions both of refractive index n', where $n > n'$. The propagation of light in an optical fiber waveguide consisting of a core and concentric cladding is similar to the propagation of light in the slab for those rays that pass through the axis of the fiber (meridional rays) so that analysis of a slab guide is a useful indication of the behavior of optical fibers. A ray of light may readily propagate down a step index fiber in which $n > n'$ as a consequence of total internal reflection for angles of incidence at the core, $i \geq i_c$, where i_c is the critical angle and

$$i_c = \sin^{-1}\left(\frac{n'}{n}\right).$$

(5.15)

There will be an infinite number of such rays with different angles of incidence at the core boundary at angles greater than the critical angle i_c given by Eq. 5.15 that may propagate along the fiber.

However, there will be interference effects between rays due to differences in optical paths so that only rays at certain angles of incidence can be propagated. In Figure 5.4 a ray ABC is shown in the waveguide. There are other rays parallel to it and a wavefront $FEAC$ is shown (a plane of constant phase). There is destructive interference in cases where the phase at C differs from that at A by values other than a multiple of 2π. The phase difference depends upon the distance $AB+BC$ and the phase change on reflection at B. The polarization commonly used in optical fibers has the electric field perpendicular to the plane of incidence thus the phase change on reflection is given by Eq. 5.13.

We require the condition

$$(AB + BC)\frac{2\pi n}{\lambda} - 2\phi = 2\pi m, \tag{5.16}$$

where m is an integer, for wavelength λ to be propagated at angles of incidence i to the surface of the waveguide. The negative sign arises here because the phase further along the fiber will be decreased and the phase is increased upon reflection. From the diagram in Figure 5.4 for the triangle ABC,

$$AB = BC \cos 2i, \tag{5.17}$$

and hence

$$(AB + BC) = BC(cos2i + 1). \tag{5.18}$$

Since

$$\cos 2i = 2 \cos^2 i - 1, \tag{5.19}$$

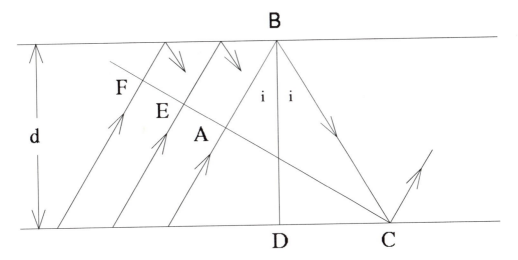

Figure 5.4: Cross-section of an optical waveguide.

then

$$AB + BC = 2BC \cos^2 i. \tag{5.20}$$

From triangle BDC,

$$BC \cos i = d; \tag{5.21}$$

so that

$$AB + BC = 2d \cos i. \tag{5.22}$$

Substituting in Eq. 5.16 gives

$$\frac{4\pi nd \cos i}{\lambda} - 2\phi = 2\pi m \tag{5.23}$$

or

$$\frac{2\pi nd \cos i}{\lambda} - \phi = \pi m. \tag{5.24}$$

For each value of m there will be a corresponding value of i, the angle i_m, that satisfies Eq. 5.24. It is not possible to obtain explicit solutions to Eq. 5.24 because ϕ depends upon i, through Eq. 5.13. Rearranging Eq. 5.24 to help find values for i_m,

$$\cos i_m = \frac{(m\pi + \phi)\lambda}{2\pi dn}. \tag{5.25}$$

Remember that i_m can only take values in the range i_c to $\frac{\pi}{2}$ and from Eq. 5.25 small values of m correspond to large values of i_m, which means that i_c determines the maximum value of m.

In the program RAYTRACE, values for i_m are calculated iteratively using an initial approximate value for ϕ and then re-calculating i_m by correcting ϕ from Eq. 5.13.

Thus it is seen that there are a finite number of angles at which rays can propagate along a particular waveguide. Because of the different ray angles the different rays will traverse the same length of guide in different times. This is known as mode dispersion, where the allowed rays are called modes. In a long communications fiber this means that a single entrant pulse will broaden and possibly separate out into a number of pulses. This could lead to confusion of signals. Mode dispersion is a serious limitation on the design and use of optical fibers. Fortunately, it can be overcome by restricting the allowed modes or by reducing the rate at which pulses are sent along the fiber. Since most optical fibers are used for digital communications rather than analogue communications, the degradation of the pulses does not affect the quality of the data transfer in well designed systems.

In the program RAYTRACE, the fiber parameters can be varied and the modal dispersion is illustrated for a chosen input pulse. In the screen display in Figure 5.5 the **Fiber** option of RAYTRACE is shown. The sliders can be adjusted to investigate

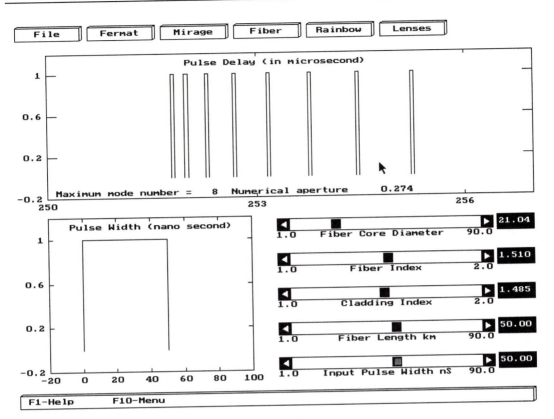

Figure 5.5: The **Fiber** option.

a wide range of parameter values. The graph at the top automatically selects the relevant region for the display.

In practical optical communications fibers in which more than one propagation mode is used there will be modal conversion along the fiber because the value of i will vary due to bends in the fiber. This intermodal conversion means that a single entrant pulse will not emerge as a train of pulses but as a single broadened pulse with an amplitude variation depending upon the bends in the route followed by the fiber. The modal conversion can be beneficial as it may help to increase the rate at which pulses may be sent along a multimode fiber, since it tends to reduce pulse broadening.

5.5.1 Numerical Aperture of the Optical Fiber

The value of α, the semi-angle of the cone of rays that can enter the fiber, is determined from Snell's law and the condition for total internal reflection in Eq. 5.15. Referring to the angles shown in Figure 5.6, for a fiber with a core of refractive index n_1 and cladding of refractive index n_2, in air of refractive index n_0, Snell's law gives

$$n_0 \sin \alpha = n_1 \sin (90 - i_c). \tag{5.26}$$

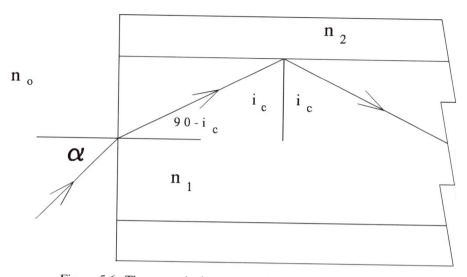

Figure 5.6: The numerical aperture of an optical waveguide, α.

Thus,

$$n_0 \sin \alpha = n_1 \cos\left[\sin^{-1}\left(\frac{n_2}{n_1}\right)\right], \tag{5.27}$$

or

$$n_0 \sin \alpha = \sqrt{(n_1)^2 - (n_2)^2}. \tag{5.28}$$

The numerical aperture, NA, defined as $n_0 \sin \alpha$, is usually specified for an optical fiber rather than α, since NA is independent of the source refractive index, n_0, as is seen in Eq. 5.28, and depends only on the refractive indices of the core, n_1, and that of the concentric cladding, n_2. This value determines the amount of light that can enter the fiber.

5.6   Formation of the Rainbow

The most common situation in which a rainbow is seen is when a shower of rain is illuminated by sunlight behind the observer. Rainbows can also be seen in clouds of drops formed in other ways if they are illuminated in a similar manner. The sequence of colors in the single bright arc of the primary rainbow is invariably violet as the innermost color, blurring through indigo, blue, green, yellow, and orange to the outermost color, red. Sometimes a secondary rainbow will be seen in an arc above and around the primary rainbow with the color sequence reversed. Other features, such as supernumerary rainbows (faint fringes below the main arc) may occasionally be seen. They cannot be explained by simple ray tracing because they are the result of interference effects[9,11] and will not be discussed further here.

Rainbows are observed, since there is a minimum in the angle through which a raindrop bends a ray of light, which depends on the distance of the ray from the center-line of the raindrop and on the variation of refractive index with wavelength.

This distance from the center-line of the drop is called the impact parameter. At the minimum angle the intensity is increased because many rays are deviated by nearly the same amount, and as a consequence of the variation in the refractive index with wavelength, the colors of the spectrum have their maximum intensities at different angles.

In Figure 5.7 a ray is shown entering a raindrop of radius r at impact parameter t. The entrant ray produces a number of reflected and refracted rays. The rays emerging from the raindrop are numbered sequentially. Rays of class 1 are directly reflected by the drop and those of class 2 are directly transmitted through it. Those that produce the primary and secondary rainbow arcs are the rays of classes 3 and 4, respectively, which suffer internal reflection once and twice, respectively. For each class of scattered rays the deviation varies over a wide range of values as a function of the impact parameter. Since in sunlight the raindrops are illuminated at all impact parameters simultaneously, light is scattered in almost all directions. An observer will see rays emerging from different raindrops at the same angle for a particular value of the impact parameter, with all the rays of one color seen at the same angle from the observer's straight-ahead line of view, thus giving the familiar circular arc of the rainbow. A full rainbow would be a circle formed by a cone of rays directed towards the observer's eye from infinity. There is no end to a rainbow and hence no pot of gold! The paths of rays through a raindrop and the variation in deviation of the rays are shown in the program RAYTRACE.

The screen in Figure 5.8 shows the **Rainbow** option display for one particular value of the impact parameter. The user can adjust the slider to investigate the variation of the deviation as a function of the impact parameter. The ray diagram is updated to show the current ray paths.

The angular spread of the rainbow was first calculated by Newton. If all the incident illumination is regarded as a bundle of parallel rays coming from one direction, the angular extent would be $1°45'$. If the angular diameter of the

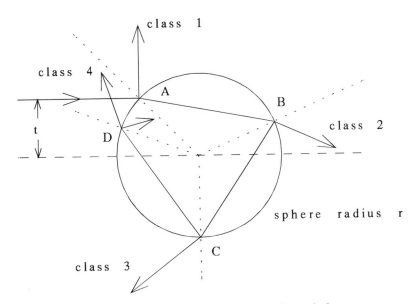

Figure 5.7: The path of a ray of light through a raindrop.

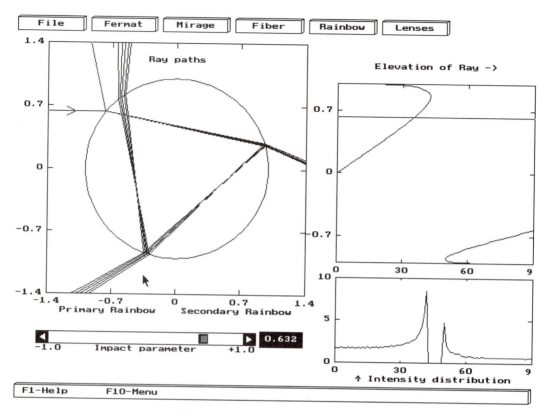

Figure 5.8: The **Rainbow** option.

sun—about half a degree—is included, Newton calculated that the total angular width would be $2^o 15'$, which he found to be in good agreement with observation.

The article by Berry and Howls[9] includes a photograph of a rainbow taken at Newton's birthplace in England and the article by Nussenzveig[11] shows a rainbow at Johnstone Straight in British Columbia. Careful examination of the photographs reveals features other than those which can be explained by simple ray-tracing ideas. There are supernumerary arcs which arise due to interference effects. The interested reader is referred to the article by Berry and Howls[9] and the article by Nussenzveig.[11]

5.7 Refraction at the Spherical Surface of a Lens

Consider the spherical refracting surface shown in Figure 5.9. The refractive index is n_1 to the left of the spherical surface and n_1' to the right of it. The center of curvature of the spherical surface, C, of radius, r_1, is on the optic axis $OACI$, which is the axis of central symmetry of the system. The point A where the spherical surface cuts the optic axis is called the vertex of the spherical surface. The path of a ray from the object at point O on the optic axis to the image at point I on the

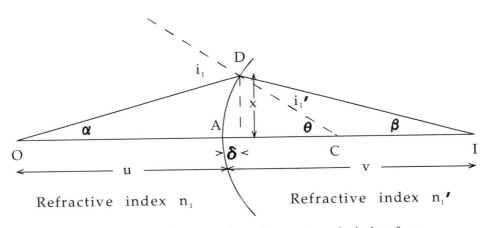

Figure 5.9: The refraction of a ray at a spherical surface.

optic axis is shown passing through a point D on the spherical refracting surface at height x above the optic axis. A radial line from the center of curvature, C, is shown passing through D. This line is the local normal to the surface at D. The angle of incidence of the ray, i_1, is the angle between the ray OD and the normal to the surface at D. The angle of refraction of the ray, i_1', is the angle between the normal to the surface and the refracted ray DI. From Snell's law,

$$n_1 \sin i_1 = n_1' \sin i_1'. \tag{5.29}$$

The application of this equation forms the basis for all ray tracing for lens design.

5.8 The Paraxial Approximation

In the derivation of the equations that give the paths of rays traveling through spherical refracting surfaces, the following conventions are adopted. The rays of light travel from left to right. The distances will be thought of as positive when measured from left to right. Distances will always be measured along the optic axis from the vertices of refracting surfaces. Distances of points above the optic axis will be taken as positive and those distances of points below the optic axis will be taken as negative.

The equations that are derived are simplified by considering only those rays close to the optic axis. In this paraxial approximation two assumptions are made. In the first it is assumed that in Figure 5.9 the plane of D is identical to the plane of the vertex of the spherical surface corresponding to $\delta = 0$. The second assumption is that for rays close to the axis, the angles i_1 and i_1' will be small. Thus we can use the approximations $\sin i \approx i$, $\cos i \approx 1$ and

$$\tan i \approx i \tag{5.30}$$

(i is in radians) using the first terms of the Taylor expansions for $\sin i$ and $\cos i$.

Thus, in the paraxial approximation for the ray ODI, we let the angle DOA be α, the angle DIA be β, the angle DCA be θ, the distance of the object point from the refracting surface be u_1, the distance of the image of the object from the refracting surface be v_1, and the height of the ray at the refracting surface be x. Then from the external angle theorem,

$$\alpha + \theta = i_1 \tag{5.31}$$

$$\theta = \beta + i_1' \tag{5.32}$$

and

$$n_1 i_1 = n_1' i_1'. \tag{5.33}$$

From Eqs. 5.31–5.33,

$$n_1(\alpha + \theta) = n_1'(\theta - \beta) \tag{5.34}$$

or

$$n_1\alpha + n_1'\beta = (n_1' - n_1)\theta. \tag{5.35}$$

Remembering that

$$\tan(\alpha) = \frac{x}{u_1} \tag{5.36}$$

$$\tan(\beta) = \frac{x}{v_1} \tag{5.37}$$

$$\tan(\theta) = \frac{x}{r_1} \tag{5.38}$$

and using the approximation for small angles, we get $\alpha = x/u_1$, $\beta = x/v_1$ and $\theta = x/r_1$. Now on substituting into Eq. 5.35 we get

$$\frac{n_1}{u_1} + \frac{n_1'}{v_1} = \frac{n_1' - n_1}{r_1}. \tag{5.39}$$

Note that the position of the image of the object produced by the refracting surface is independent of x for paraxial rays. Thus all rays from the object that are close to the axis, paraxial rays, meet again at the image point I. The quantity $P = \left(n_1' - n_1\right)/r_1$ is called the power of the surface. With r_1 measured in meters, the unit of power is the dioptre. For a lens in air $n_1 = 1$ and

$$P_1 = \frac{n_1' - 1}{r_1}. \tag{5.40}$$

5.8.1 Thin Lens Equation

A lens consists of two spherical refracting surfaces, of separation t, bounding a region of refractive index n in air of refractive index 1. Equation 5.39 is perfectly general, and it can be applied at the second surface of the lens shown in Figure 5.10, so that

$$\frac{n_2}{u_2} + \frac{n_2'}{v_2} = \frac{n_2' - n_2}{r_2}. \tag{5.41}$$

The image produced by the first refracting surface acts as the object for the second refracting surface and clearly for a thin lens of thickness t with $t = 0$, $u_2 = -v_1$, $n_2 = n_1'$, and $n_2' = n_1$, as shown in Figure 5.10. Adding Eqs. 5.39 and 5.41,

$$\frac{n_1}{u_1} + \frac{n_1'}{v_1} - \frac{n_1'}{v_1} + \frac{n_1}{v_2} = \frac{n_1' - n_1}{r_1} + \frac{n_2' - n_2}{r_2}. \tag{5.42}$$

Therefore,

$$\frac{n_1' - n_1}{r_1} - \frac{n_1' - n_1}{r_2} = \frac{n_1}{u_1} + \frac{n_1}{v_2}, \tag{5.43}$$

and for a thin lens in air with an object at distance u and the image at distance v (the subscripts to u and v are omitted since the lens surfaces are at the same position on the optic axis if $t = 0$),

$$\frac{1}{u} + \frac{1}{v} = (n - 1)\left(\frac{1}{r_1} + \frac{1}{r_2}\right), \tag{5.44}$$

The image distance for an object at infinity is called the focal length f; and

$$\frac{1}{f} = (n - 1)\left(\frac{1}{r_1} + \frac{1}{r_2}\right). \tag{5.45}$$

It follows that

$$\frac{1}{f} = \frac{1}{u} + \frac{1}{v}. \tag{5.46}$$

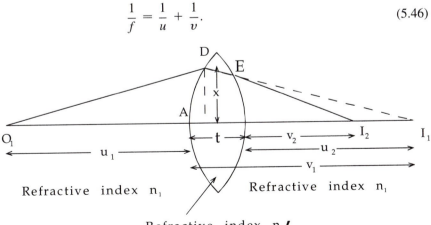

Figure 5.10: The path of a ray through a lens.

This is Gauss' form of the lens equation. The point on the axis at which the image of an object at infinity is formed is called the focal point. A lens has two focal points: for a converging lens, the first focal point is the point on the axis to the left of the lens from which rays become parallel to the axis as they emerge from the lens; the second focal point is the point on the axis at which light travelling from left to right from an object on the axis at minus infinity is brought to a focus.

The object and image distance can be measured with respect to the focal points of a lens, in which case let $u = z_1 + f$ and $v = z_2 + f$, then

$$\frac{1}{z_1 + f} + \frac{1}{z_2 + f} = \frac{1}{f}. \tag{5.47}$$

Therefore,

$$\frac{z_2 + f + z_1 + f}{(z_1 + f)(z_2 + f)} = \frac{1}{f}; \tag{5.48}$$

so that

$$z_1 z_2 = f^2, \tag{5.49}$$

which is Newton's form of the lens equation.

5.8.2   Graphical Method

The image formed by a single thin lens or spherical mirror can easily be located by graphical means. This method is often used in introductory physics courses as a supplement to the more abstract algebraic method, since it enables users to get a grasp of the concept of image formation quickly. See, for example, Tipler.[5] For a lens, there are three principal rays:

1. The *parallel* ray. This ray is drawn from the object to the lens parallel to the optic axis. At the lens it is deflected to pass through (in the case of a converging lens) or appear to come from (in the case of a diverging lens) the second focal point of the lens.
2. The *central* ray, which is drawn through the center of the lens and continues on without deflection since for a thin lens at the vertex the lens has nearly plane parallel surfaces perpendicular to the axis.
3. The *focal* ray. This ray runs through the first focal point (in the case of a converging lens) or towards the first focal point (in the case of a diverging lens) until it reaches the lens. It is then deflected to run parallel to the optic axis.

If a real image is formed, these three rays cross one another at the image point. Otherwise, they appear to diverge from the position of the virtual image. It is easy to show from this construction that the object and image distances, u and v, satisfy the thin lens (see Eq. 5.46).

The screen display shown in Figure 5.11 was produced by the auxilliary program TWOLENS. This figure shows the initial ray diagram for a lens. The object is represented by the vertical arrow on the left and the image by the vertical arrow on the right. The three principal rays are represented by the horizontal and sloping lines proceeding from the object via the lens toward an observer. The lens is represented by the vertical line in the middle of the figure and its focal points are represented by the two short lines on the optic axis.

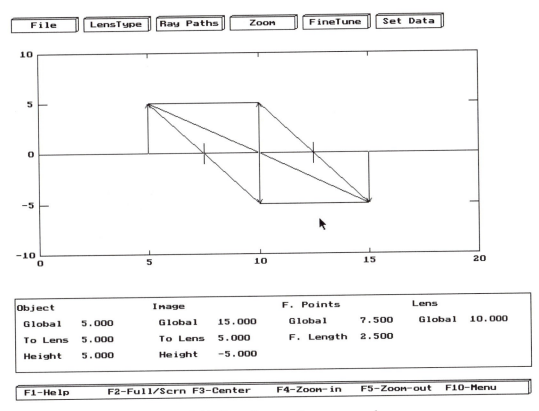

Figure 5.11: Ray diagram for a convex lens.

5.9   The Matrix Method—for Paraxial Rays

In following the passage of paraxial rays of light through a series of thick lenses ($t \neq 0$) it is convenient to use the matrix notation to describe the ray, as well as the refraction and translation of the ray. In paraxial optics two coordinates are sufficient to describe a ray at any position along the axis of the optical system. The two coordinates are the distance x of the ray from the axis, and the angle α the ray makes to the axis. The coordinates that are used in the matrix method are x and λ where $\lambda = n\alpha$ and n is the local refractive index.

Consider the first surface of a lens of refractive index n_1' in air of refractive index n_1, shown in Figure 5.10. For the refraction at D, Eq. 5.39 is used. Let the first spherical refracting surface have a radius r_1. Let the distance of the ray from the optic axis before refraction at D be x_1 and let it make angle α_1 to the optic axis. After refraction let the distance of the ray from the optic axis be x_1' and let the angle it makes to the positive optic axis be α_1'. In Figure 5.9 it can be seen that $-\beta = \alpha_1'$, where β is the angle the refracted ray makes with the axis.

Clearly, at D

$$x_1' = x_1. \tag{5.50}$$

Combining Eqs. 5.35, 5.38, and 5.50, the result in matrix format is

$$\begin{pmatrix} \lambda_1' \\ x_1' \end{pmatrix} = \begin{pmatrix} 1 & -P_1 \\ 0 & 1 \end{pmatrix} \begin{pmatrix} \lambda_1 \\ x_1 \end{pmatrix}, \tag{5.51}$$

where $\lambda_1' = n_1'\alpha_1'$ and $\lambda_1 = n_1\alpha_1$ with $P_1 = \left(n_1' - n_1\right)/r_1$. The matrix

$$\begin{pmatrix} 1 & -P_1 \\ 0 & 1 \end{pmatrix}$$

is called the refraction matrix $\mathbf{R}_1$. The passage of a ray through an optical system can be described in terms of the ray coordinates λ and x at various positions along the optical axis. In Eq. 5.51 the ray coordinates after refraction by a spherical surface are given in terms of the product of the refraction matrix and the coordinates before refraction. The other process that occurs for a ray is that it travels from one position to another as shown in Figure 5.10, where the ray travels from the first refracting surface to the second refracting surface between D and E. The distance along the optical axis between D and E, for paraxial rays, is the thickness of the lens t. The ray coordinates inside the lens at D are λ_1' and x_1', and the ray coordinates at E inside the lens are λ_2 and x_2. The ray between D and E travels through a homogeneous refracting material of index n_1', and thus the ray travels along a straight line at a constant angle to the optical axis so that $\lambda_2 = \lambda_1'$ and the change in distance of the ray from the optical axis is given by

$$x_2 - x_1' = t \tan\left(\alpha_1'\right), \tag{5.52}$$

and for small angles

$$x_2 - x_1' = t\alpha_1' = \lambda_1'\frac{t}{n}. \tag{5.53}$$

The ray coordinates at E, before refraction, can thus be given in terms of the ray coordinates at D multiplied by the matrix

$$\begin{pmatrix} 1 & 0 \\ t/n_1' & 1 \end{pmatrix}$$

which is called the translation matrix $\mathbf{T}$. In matrix notation

$$\begin{pmatrix} \lambda_2 \\ x_2 \end{pmatrix} = \begin{pmatrix} 1 & 0 \\ t/n_1' & 1 \end{pmatrix}\begin{pmatrix} \lambda_1' \\ x_1' \end{pmatrix} \tag{5.54}$$

To complete the description of the passage of the ray through the thick lens in Figure 5.10 it is necessary to obtain the angle the ray makes with the optic axis after refraction by the second surface at E, α_2'. The distance of the ray from the optic axis, x_2', is the same as the distance before refraction, x_2. The matrix equation for the refraction is

$$\begin{pmatrix} \lambda_2' \\ x_2' \end{pmatrix} = \begin{pmatrix} 1 & -P_2 \\ 0 & 1 \end{pmatrix}\begin{pmatrix} \lambda_2 \\ x_2 \end{pmatrix}, \tag{5.55}$$

where $P_2 = (n_2' - n_2)/r_2$, r_2 is the radius of curvature of the second surface, $n_2 = n_1'$ and $n_2' = n_1$.

5.9.1   The System Matrix

The passage of the ray through the lens from before the first refracting surface to after the second refracting surface is given by the equation

$$\begin{pmatrix} \lambda_2' \\ x_2' \end{pmatrix} = \begin{pmatrix} 1 & -P_2 \\ 0 & 1 \end{pmatrix} \begin{pmatrix} 1 & 0 \\ t/n_1 & 1 \end{pmatrix} \begin{pmatrix} 1 & -P_1 \\ 0 & 1 \end{pmatrix} \begin{pmatrix} \lambda_1 \\ x_1 \end{pmatrix} \tag{5.56}$$

or more compactly

$$\begin{pmatrix} \lambda_2' \\ x_2' \end{pmatrix} = \mathbf{R_2 T R_1} \begin{pmatrix} \lambda_1 \\ x_1 \end{pmatrix}, \tag{5.57}$$

where $\mathbf{R_1}$ and $\mathbf{R_2}$ are the refraction matrices for the first and second surfaces, respectively, and $\mathbf{T}$ is the translation matrix from the first to the second surface. The product $\mathbf{R_2 \, T \, R_1}$ is a single $\mathbf{2 \times 2}$ matrix. Compound lenses consisting of several refracting surfaces separated by distances $t_1, t_2 \ldots. t_n$ can be treated in a similar way and the ray coordinates of a ray from before the first refracting surface of a compound lens to after the last refracting surface can be calculated by use of a single matrix formed by the product of the separate refraction and translation matrices. The resultant $\mathbf{2 \times 2}$ matrix is called the system matrix $\mathbf{S}$, and so for the lens in Figure 5.10

$$\mathbf{S} = \mathbf{R_2 T R_1}. \tag{5.58}$$

The system matrix may involve extra translation matrices if a plane, other than that of the first refracting surface, is used to define the start or end of the lens, or optical system. It is interesting to note that the determinants of the refraction and translation matrices are unity and, as a consequence of the properties of matrices, the determinant of the system matrix is also unity.

The coefficients of the system matrix are called the Gaussian constants of the optical system and

$$\mathbf{S} = \begin{pmatrix} \mathbf{b} & -\mathbf{a} \\ -\mathbf{d} & \mathbf{c} \end{pmatrix}; \tag{5.59}$$

so that for the lens in Figure 5.10 the coefficients are obtained from Eq. 5.53 as

$$a = P_1 + P_2 - P_1 P_2 \frac{t}{n_1'} \tag{5.60}$$

$$b = 1 - P_2 \frac{t}{n_1'} \tag{5.61}$$

$$c = 1 - P_1 \frac{t}{n_1'} \tag{5.62}$$

$$d = -\frac{t}{n_1'}. \tag{5.63}$$

5.9.2   Principal Planes

The two focal points of a lens are two of the six cardinal points of the lens system. The planes perpendicular to the optic axis through the focal points are called the focal planes. The location of the focal points for a thick lens can be found in terms of the coefficients of the system matrix, which in turn are determined by the curvature and position of the various spherical surfaces that separate the materials of different refractive index. Consider a ray from an object on the axis at infinity with coordinates (λ_0, x_0) and let the coordinates of the same ray at the focal point be (λ_i, x_i); then

$$\begin{pmatrix} \lambda_i \\ x_i \end{pmatrix} = \mathbf{T_i S T_0} \begin{pmatrix} \lambda_0 \\ x_0 \end{pmatrix}, \tag{5.64}$$

where $\mathbf{T_0}$ is the translation matrix to bring the ray from infinity to before the first surface of the lens and $\mathbf{T_i}$ is the image translation matrix to take the ray from after the second (last) surface of the lens to the image point. Therefore

$$\begin{pmatrix} \lambda_i \\ x_i \end{pmatrix} = \begin{pmatrix} 1 & 0 \\ v_f & 1 \end{pmatrix} \begin{pmatrix} b & -a \\ -d & c \end{pmatrix} \begin{pmatrix} 1 & 0 \\ \infty & 1 \end{pmatrix} \begin{pmatrix} 0 \\ x_0 \end{pmatrix}, \tag{5.65}$$

where v_f is the distance from the second surface of the lens to the focal point, and since the rays from an object on the axis at infinity are parallel to the axis, then $\lambda_0 = 0$. Multiplying the last three matrices together gives

$$\begin{pmatrix} \lambda_i \\ x_i \end{pmatrix} = \begin{pmatrix} 1 & 0 \\ v_f & 1 \end{pmatrix} \begin{pmatrix} -ax_0 \\ cx_0 \end{pmatrix}. \tag{5.66}$$

At the focal point $x_i = 0$; with $\lambda_i \neq 0$; then

$$0 = v_f(-ac_0) + (cx_0). \tag{5.67}$$

Hence,

$$v_f = \frac{c}{a}. \tag{5.68}$$

Similarly it can be shown that

$$u_f = \frac{b}{a}, \tag{5.69}$$

where u_f is the distance of the first focal point from the first surface of the lens.

5.9.3   Newton's Formula for a Thick Lens

Once the focal points have been located, object and image distances can be measured to the focal points. The distance of the object from the first surface is given by $u_s = z_1 + u_f$ and the distance of the image from the second surface is given by $v_s = z_2 + v_f$. Let a ray from the object point have coordinates (λ_0, x_0) and let the coordinates for the same ray at the image point be (λ_i, x_i); then

$$\begin{pmatrix} \lambda_i \\ x_i \end{pmatrix} = \begin{pmatrix} 1 & 0 \\ z_2 + \frac{c}{a} & 1 \end{pmatrix} \begin{pmatrix} b & -a \\ -d & c \end{pmatrix} \begin{pmatrix} 1 & 0 \\ z_1 + \frac{b}{a} & 1 \end{pmatrix} \begin{pmatrix} \lambda_0 \\ x_0 \end{pmatrix}. \tag{5.70}$$

Multiplying out gives

$$\begin{pmatrix} \lambda_i \\ x_i \end{pmatrix} = \begin{pmatrix} 1 & 0 \\ z_2 + \frac{c}{a} & 1 \end{pmatrix} \begin{pmatrix} b\lambda_0 - a\left(z_1 + \frac{b}{a}\right)\lambda_0 - ax_0 \\ -d\lambda_0 + c\left(z_1 + \frac{b}{a}\right)\lambda_0 + cx_0 \end{pmatrix}. \tag{5.71}$$

At the image point $x_i = 0$ for $x_0 = 0$; thus, the x coordinate of the ray gives

$$0 = \left(z_2 + \frac{c}{a}\right)\left(b\lambda_0 - a\left(z_1 + \frac{b}{a}\right)\lambda_0 - ax_0\right)$$
$$-d\lambda_0 + c\left(z_1 + \frac{b}{a}\right)\lambda_0 + cx_0. \tag{5.72}$$

Expanding, and remembering that det $\mathbf{S} = 1$ so that $bc - ad = 1$, yields

$$0 = -az_1z_2 - \frac{1}{a}. \tag{5.73}$$

Thus

$$z_1z_2 = \frac{1}{a^2}. \tag{5.74}$$

Hence, comparing this result with Eq. 5.49, Newton's formula for a thin lens, it is seen that $1/a = f$, the focal length of the lens. In the program RAYTRACE the paraxial approximation is used to calculate the focal length of the lens and the position of the focal plane. The distance of the focal plane from the last surface of the lens, v_f, is called the back focal distance.

5.9.4   Gauss' Formula for a Thick Lens

It is convenient to be able to use Gauss' formula, Eq. 5.46, to describe the behavior of thick lenses. Consider again the thick lens shown in Figure 5.10. The positions of the focal points are given by Eqs. 5.68 and 5.69.

In Gauss' formula the object and image distance are measured to the plane of the thin lens. For a thick lens it is no longer sensible to talk of "the plane of the lens." By comparing the behavior of a thick lens with that of a thin lens, it is reasonable to suggest that the object and image distance should be measured from two planes U_1 and U_2, such that these two planes are at a distance equivalent to the focal length from the focal points F_1 and F_2 as shown in Figure 5.12. It is possible to relate the distance, u, of the object from the plane U_1 to the distance v, of the image from the plane U_2 using the matrix notation. The planes U_1 and U_2 are called the unit planes for reasons that will become apparent. The distance of U_1 from the first surface of the lens is $(1 - b)/a$, since the first focal plane is at distance b/a from the first surface of the lens and since $1/a = f$ from Eq. 5.74. Similarly, it may be seen that the distance of the second unit plane from the second surface of the lens is $(1 - c)/a$. The distance of the object from the first plane of the lens u_1 and the distance of the image from the second (or last) plane is v_2. Thus,

$$u = u_1 + \frac{1 - b}{a} \tag{5.75}$$

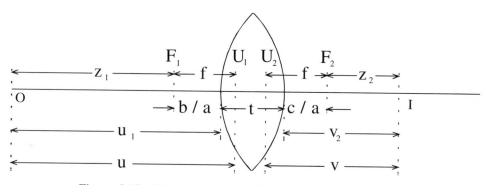

Figure 5.12: Object and image distances for a thick lens.

$$v = v_2 + \frac{1 - c}{a}, \tag{5.76}$$

and thus

$$u_1 = u + \frac{b - 1}{a} \tag{5.77}$$

$$v_2 = v + \frac{c - 1}{a}. \tag{5.78}$$

Substituting into Eq. 5.70 gives

$$\begin{pmatrix} \lambda_i \\ x_i \end{pmatrix} = \begin{pmatrix} 1 & 0 \\ v + \frac{c-1}{a} & 1 \end{pmatrix} \begin{pmatrix} b & -a \\ -d & c \end{pmatrix} \begin{pmatrix} 1 & 0 \\ u + \frac{b-1}{a} & 1 \end{pmatrix} \begin{pmatrix} \lambda_0 \\ x_0 \end{pmatrix}, \tag{5.79}$$

which becomes

$$\begin{pmatrix} \lambda_i \\ x_i \end{pmatrix} = \begin{pmatrix} 1 & 0 \\ v + \frac{c-1}{a} & 1 \end{pmatrix} \begin{pmatrix} b\lambda_0 - a\left(u\lambda_0 + \left(\frac{b-1}{a}\right)\lambda_0 + x_0\right) \\ -d\lambda_0 + c\left(u\lambda_0 + \left(\frac{b-1}{a}\right)\lambda_0 + x_0\right) \end{pmatrix}. \tag{5.80}$$

For an object on the axis ($x_0 = 0$) and considering rays with $\lambda_0 \neq 0$, for the image point it is required that $x_i = 0$, and thus,

$$0 = vb\lambda_0 - av\left(u\lambda_0 + \left(\frac{b - 1}{a}\right)\lambda_0\right) + \left(\frac{c - 1}{a}\right)b\lambda_0 - \left(\frac{c - 1}{a}\right)$$
$$\cdot a\left(u\lambda_0 + \left(\frac{b - 1}{a}\right)\lambda_0\right) - d\lambda_0 + c\left(u\lambda_0 + \left(\frac{b - 1}{a}\right)\lambda_0\right). \tag{5.81}$$

This simplifies to

$$0 = -avu + v + u, \tag{5.82}$$

which can be rearranged as

$$\frac{1}{u} + \frac{1}{v} = a = \frac{1}{f}, \tag{5.83}$$

which is the Gauss form of the lens equation for a thick lens with object distance and image distance measured from the first and second unit planes, respectively.

5.9.5   Magnification

The ratio of the size (distance of a point from the optic axis) of an image point to the size of the corresponding object point is referred to as the magnification. Consider an object at distance u_1 from the first surface of the lens that has an image at distance v_2 from the second (or last) surface. Substituting into Eq. 5.70 gives

$$\begin{pmatrix} \lambda_i \\ x_i \end{pmatrix} = \begin{pmatrix} 1 & 0 \\ v_2 & 1 \end{pmatrix} \begin{pmatrix} b & -a \\ -d & c \end{pmatrix} \begin{pmatrix} 1 & 0 \\ u_1 & 1 \end{pmatrix} \begin{pmatrix} \lambda_0 \\ x_0 \end{pmatrix}, \tag{5.84}$$

which can be expanded to

$$\begin{pmatrix} \lambda_i \\ x_i \end{pmatrix} = \begin{pmatrix} b - au_1 & -a \\ bv_2 - au_1v_2 + cu_1 - d & c - av_2 \end{pmatrix} \begin{pmatrix} \lambda_0 \\ x_0 \end{pmatrix}. \tag{5.85}$$

As already discussed for an object on the axis ($x_0 = 0$), considering rays with $\lambda_0 \neq 0$ and since at the image point x_i is equal to zero, it follows that

$$bv_2 - au_1v_2 + cu_1 - d = 0. \tag{5.86}$$

Then Eq. 5.85 can be rewritten as

$$\begin{pmatrix} \lambda_i \\ x_i \end{pmatrix} = \begin{pmatrix} b - au_1 & -a \\ 0 & c - av_2 \end{pmatrix} \begin{pmatrix} \lambda_0 \\ x_0 \end{pmatrix}. \tag{5.87}$$

As this equation will still apply for points on the object at which $x_i \neq 0$, then, considering the x term only, Eq. 5.87 gives

$$x_i = (c - av_2)x_0; \tag{5.88}$$

so that the magnification, M, is in this case given by

$$M = \frac{x_i}{x_0} = c - av_2. \tag{5.89}$$

Now remember that the determinants of the various matrices used are all equal to one and that thus the determinant of their products will be one. Then from Eq. 5.89

$$\det \begin{pmatrix} b - au_1 & -a \\ 0 & c - av_2 \end{pmatrix} = 1; \tag{5.90}$$

so that

$$b - au_1 = \frac{1}{c - av_2}. \tag{5.91}$$

It is possible to rewrite Eq. 5.87 in terms of the magnification ratio so that

$$\begin{pmatrix} \lambda_i \\ x_i \end{pmatrix} = \begin{pmatrix} \frac{1}{M} & -a \\ 0 & M \end{pmatrix} \begin{pmatrix} \lambda_0 \\ x_0 \end{pmatrix}. \tag{5.92}$$

The unit planes, as defined above, have been fixed in terms of their distance from the focal points.

The rays from object to image, for $\lambda_0 \neq 0$ and $x_0 \neq 0$, can be followed by the matrix method between any two planes. Suppose now that the height of a ray at the first unit plane is x_1 and that its height at the second unit plane is x_2; then substituting the distance of the unit planes into Eq. 5.87 in place of object and image distances gives

$$\begin{pmatrix} \lambda_2 \\ x_2 \end{pmatrix} = \begin{pmatrix} b - a\left(\frac{b-1}{a}\right) & -a \\ 0 & c - a\left(\frac{c-1}{a}\right) \end{pmatrix} \begin{pmatrix} \lambda_1 \\ x_1 \end{pmatrix}, \tag{5.93}$$

which simplifies to

$$\begin{pmatrix} \lambda_2 \\ x_2 \end{pmatrix} = \begin{pmatrix} 1 & -a \\ 0 & 1 \end{pmatrix} \begin{pmatrix} \lambda_1 \\ x_1 \end{pmatrix}, \tag{5.94}$$

and comparing this result with Eq. 5.92, it is seen that the magnification for rays between the unit planes is unity and therefore $x_1 = x_2$.

5.9.6   Nodal Planes

Examining the λ term in Eq. 5.94, it is seen that

$$\lambda_2 = \lambda_1 - ax_1; \tag{5.95}$$

so that for $x_1 = 0$ (and $x_2 = 0$) the result $\lambda_2 = \lambda_1$ is obtained, showing that a ray crossing the second unit plane is in the same direction as when it crossed the first unit plane. This unit angular magnification is also identified with special points and planes in the optical system called the nodal points and planes. The unit plans are coincident with the nodal planes for systems in which the final refractive index is equal to the initial refractive index, such as lenses in air which have been considered here. The six cardinal points of a lens system are the two focal points, the two points at which the unit planes cut the axis, and the two points at which the nodal planes cut the axis.

5.10   Matrix Method—for Non-Paraxial Rays

Most lens systems produce imperfect images of finite objects. The defects are called aberrations. There are three main types of aberration: chromatic aberration due to the variation of refractive index with wavelength in the glasses used; geometrical aberrations, which occur because of the use of spherical surfaces and other geometrical factors; and finally there are aberrations which arise due to the wave nature of light and produce images which are not sharp due to diffraction. In the program RAYTRACE the geometrical aberrations are illustrated and the other effects are not considered.

An important consequence of the paraxial approximation that has been used in Eqs. 5.31 to 5.95 is that all the rays from a point on the object intersect at a point in the image and thus a geometrically perfect image of the object is expected.

If non-paraxial rays are considered, then the approximations in Eq. 5.30 are no longer valid and additional terms in the Taylor series expansion need to be taken into account.

The result of considering rays from an object on the axis with angles of incidence greater than 5° to the normal to the lens surface is that the distance of best focus (intersection of the rays) will depend on the region of the lens through which the rays pass. This absence of a single position of best focus leads to a blurred image, which is the result of the geometry of the passage of the rays and is thus a geometrical aberration. In particular, for meridional rays (those passing at some point through the axis) it is a consequence of the use of spherical refracting surfaces and for this reason is called spherical aberration. There are two ways of expressing this aberration, which are either as longitudinal spherical aberration or as transverse spherical aberration. Other aberrations due to the imperfect imaging of spherical lens systems are found to occur when off-axis objects points are considered. The rays from off-axis points, which do not pass through the axis, are called skew rays.

In order to follow the true paths of rays through lenses it is necessary to abandon the paraxial approximation. Non-paraxial rays are considered in detail by Nussbaum and Phillips,[6] who give the equations for an exact ray trace. The point objects of paraxial optics are replaced by finite objects, so that this ray tracing is often called a finite ray trace. In the program RAYTRACE an exact ray trace is used for the rays from point objects to show the image defects that occur for non-axial rays.

5.11   Monochromatic Aberrations in Lenses

A lens system is designed to reduce the effects of the aberrations that are significant in the intended use of the lens usually at the cost of a worse performance for other aberrations. As shown in Eq. 5.70 using the paraxial analysis, the height of the image above the axis, x', is given in terms of the height of the object above the axis x by

$$\begin{pmatrix} \lambda' \\ x' \end{pmatrix} = \begin{pmatrix} b & -a \\ -d & c \end{pmatrix} \begin{pmatrix} \lambda \\ x \end{pmatrix}, \tag{5.96}$$

where $\lambda' = \lambda_i$ and $\mathbf{S} = \begin{pmatrix} b & -a \\ -d & c \end{pmatrix}$. The matrix $\mathbf{S}$ is the product of the three **2 × 2** matrices in Eq. 5.70 and is the matrix needed to trace any paraxial ray from the object plane to the image plane. Thus

$$\lambda' = b\lambda - ax \tag{5.97}$$

$$x' = -d\lambda + cx \tag{5.98}$$

and eliminating λ gives

$$x' = -\frac{d}{b}(\lambda' + ax) + cx, \tag{5.99}$$

which simplifies to

$$x' = -\frac{d}{b}\lambda' + \frac{x}{b} \qquad (5.100)$$

since the determinant of **S** is unity.

Welford[8] has shown that an alternative approach to the exact ray trace for meridional and skew rays, such as that of Nussbaum and Phillips,[6] can be used to determine the corrections necessary to the paraxial ray tracing equations for rays outside the paraxial region. It has been shown[7] that for these rays

$$x' = -\frac{d}{b}\lambda'_x + \frac{x}{b} + f(x, y, \lambda'_x, \lambda'_y) \qquad (5.101)$$

and

$$y' = -\frac{d}{b}\lambda'_y + \frac{y}{b} + f(y, x, \lambda'_y, \lambda'_x), \qquad (5.102)$$

where f is the correction term called the aberration function.

The function f will contain third-order and higher-order terms, since the correction is to the small-angle approximation that $\tan\alpha = \alpha$, and the next term in the Taylor series expansion of $\tan\alpha$ is the third-order term. The first-order correction will thus consist of a 20-term expression in the four parameters. The first derivation of the correction terms was made by Seidel,[1] hence the name Seidel aberrations. It has been shown[7] that the 20 third-order terms can be reduced to 5, so that the aberration function becomes

$$
\begin{aligned}
f(x, y, &x'', y'') \\
&= Px''\left(x''^2 + y''^2\right) + Qx\left(x^2 + y^2\right) + R\left(x\left(x''^2 + y''^2\right) + 2x''(xx'' + yy'')\right) \\
&\quad + (S - T)x''\left(x^2 + y^2\right) + 2Tx(xx'' + yy''),
\end{aligned} \qquad (5.103)
$$

where x'' and y'' are the ray coordinates in the exit pupil and w is the distance from the exit pupil to the image plane. Exit pupils are explained in section 5.12. This leads to

$$\sin\lambda'_x = \frac{(x' - x'')}{w} \qquad (5.104)$$

$$\sin\lambda'_y = \frac{(y' - y'')}{w}. \qquad (5.105)$$

The coefficients $P, Q, R, S,$ and T are called the third-order aberrations. Each of the aberrations has a descriptive name which is related to the form of the aberration. The five names associated with the coefficients P, R, T, S, and Q are, respectively, spherical aberration, coma, astigmatism, field curvature, and distortion. The appearance of these aberrations is described below. A detailed account of the aberrations of lens sytems can be found in the book by Welford.[8] In the program RAYTRACE the Seidel aberrations are calculated and can be displayed by use of the hot key **F6-Seidel**.

5.11.1 Spherical Aberration

Lenses and mirrors with spherical surfaces do not produce a point image of an axial point object because the position of intersection of the rays depends on the distance from the axis at which the rays pass through the lens. Usually the deviation of rays passing through the outer rim of a lens is too large, and consequently they intersect before reaching the point at which paraxial rays intersect. Thus, instead of a single position for the image there is a point at which the image is smallest, called the circle of least confusion. Spherical aberration is reduced if the deviations of the rays are spread over several surfaces, and can be minimized if the deviation for rays is the same at each surface. It is possible by the use of aspheric surfaces to eliminate axial spherical aberration for a given wavelength.

5.11.2 Coma

Spherical aberration is the cause of imperfect imaging on the axis of the lens system. The imperfection of image points slightly off the axis is called coma. The defect of coma is particularly important in astronomy, since the image of a star may look like a comet as a consequence of this aberration, hence the name coma for the aberration. Coma can be eliminated for a given object position by choosing appropriate radii of curvature. A system that is free from coma and spherical aberration is said to be aplanatic and the two conjugate points are called aplanatic points.[2]

5.11.3 Astigmatism

The rays from a point object which is not close to the axis produce an image which shows astigmatism. The rays do not come to a point focus but form two linear images separated by a short distance. For any point object in the xz plane the linear images will be parallel and perpendicular to the xz plane. The consquence of this defect is that there is no single plane of sharp focus for an object such as a wheel with spokes that has its axis on the optical axis of the lens. If the best focus for the radial spokes is chosen, then the circumferential rim of the wheel will not be in focus. The locally orthogonal directions can not be brought into focus at the same position for non-paraxial rays due to the presence of astigmatic aberration in a lens. The plane of best focus is taken to be an intermediate plane where the image of an off-axis point object forms a circle of least confusion.

5.11.4 Field Curvature

Two further related defects can be identified in the aberrations of lenses used outside the paraxial region. In a lens which is corrected for astigmatism there may not be a single plane surface of best focus for every object point, but the points of best focus may lie on a curved surface. This is the image defect called curvature of field. In photography it is important to have the image on a plane surface and a lens which is corrected for astigmatism and curvature of field is called an anastigmat.

5.11.5   Distortion

There is one further image defect which may be present. If corrections have been made for all the other aberrations described above, the image produced by a lens may still suffer from distortion. Distortion is manifest in an image by a magnification which depends on the off-axis distance of an image point. If the magnification varies with the distance of image points from the axis, then the image will not be a faithful reproduction of the object. For an object in the form of a square the image will show barrel distortion if the magnification decreases with distance from the axis, so that the corners of the square are closer to the axis than would be expected. If the magnification increases with distance of the image point from the axis the image of a square object will show pin-cushion distortion in which the corners are further from the axis than would be expected.[3]

   Distortion cannot occur in a symmetrical lens system forming a real image of the same size as the object, because the reversibilty of rays of light allows object and image to be interchanged, and thus the image cannot be larger in one position than in the other position. Symmetrical lens systems produce little distortion for any object position.

5.12   Stops, Entrance, and Exit Pupils

A stop is an opening in a screen which would otherwise prevent light from passing further through a coaxial lens system. A stop may be an actual opening, which may be adjustable (as in the aperture of a camera) or it may be the lens itself, or one lens of a lens system. For a particular object point on the axis, one opening must limit the cone of rays forming the image, and this is called the aperture stop. The opening which acts as the aperture stop may vary with changes in object position, but for most lens systems the range of object positions is limited so that there is only one aperture stop. The image of the aperture stop in the lenses to the left of it is called the entrance pupil. The image of the aperture stop in the lenses to the right of it is called the exit pupil.

   In the telescope and microscope the objective lens forms the aperture stop, and since one of the aims is to gather as much light as possible these lenses are usually made as large as possible, consistent with the other aims of the design. In the camera an adjustable aperture is used to control the amount of light falling on the photographic film in conjunction with the time for which the shutter is open. The aperture stop will affect the magnitude of the aberrations in a lens system, and if a large aperture is not needed to gather light a smaller aperture will produce better overall performance in a lens. Stops can be introduced into the lens system in the program RAYTRACE and their effects can be investigated. The use of different colors in the program RAYTRACE to identify rays passing through different regions of the lens can also be used to consider the effects of stops. A fuller account of the effects of apertures on the aberrations of lenses is given by Meyer-Arendt.[2]

5.13 Running RAYTRACE

There are six headings in the main menu of the program RAYTRACE namely: **File**, **Fermat**, **Mirage**, **Fiber**, **Rainbow**, and **Lenses**. (The **File** heading pull-down menu includes options to load and save data from the **Lenses** menu option for which data defining lens design may be needed.) The load and save lens data options are only available after one of the **Lenses** options has been selected.

- **Fermat**

  This option is designed to illustrate Fermat's principle of least path for refraction and show its agreement with Snell's law of refraction. The screen display for this option is shown in Figure 5.2. A ray of light is traced from one corner of a square region until it meets a boundary which separates material of refractive index n from material of refractive index n'. At the boundary the path of a ray to the opposite corner is shown as a dotted line, and the path of a ray obeying Snell's law is shown as a continuous line. The program calculates the optical path for the route from one corner to the opposite corner. The optical path for this route depends on the position at which the ray meets the boundary between the two regions. Graphs are drawn alongside which show the optical paths for rays to all positions along the boundary between the two regions. The total optical path from corner to corner is also shown plotted against the position at which the ray might cross the boundary. It is possible, by the use of sliders on the screen, to move the position at which the ray meets the boundary and a vertical line on each graph indicates this position.

  Below the display of the ray path the numerical values change as the sliders are moved, and it will be found that the separate calculations for the total time of travel and Snell's law show that Snell's law is in agreement with Fermat's principle as the values of $n \sin i$ and $n' \sin i'$ are the same at the position at which the minimum optical path occurs, and that at this point there is just one line below the boundary. There are sliders that make it possible to change n and n', the refractive indices of the two regions, and also the vertical position of the boundary separating the two regions. When any of these three sliders is adjusted the graphs of optical path are redrawn automatically. This simulation shows the path variation expressed in Eq. 5.4. The simulation also illustrates the conditions for total internal reflection.

- **Mirage**

  In the simulation of a mirage in the program RAYTRACE, shown in the display in Figure 5.3, use is made of the reversibility of rays of light to show which rays of light could arrive at an observer's eye after "reflection" at the hot surface of a road. The simulation takes the eye of the observer to be at a height of 1.6 m above the road. Light from the sky will illuminate the road immediately in front of the observer. This will be scattered diffusely and the surface will appear dry. At some distance from the observer the rays of light from the sky will be deviated into a curve so that, at some point, they suffer total internal reflection and the mirage of a "wet" surface is formed. A slider is provided so that the direction of the ray at the observer's eye can be varied. The rays which cause the mirage of a wet road are drawn as continuous curves and the rays that come from diffuse scattering by the road are drawn as as dotted curves. The hot key **F5-Index Var** is available to change the function setting the variation of refractive index with height above the road surface.

• **Fiber**

This option gives the user the opportunity to investigate the factors that have lead to the adoption of optical fibers around 1 to 10 μm in diameter. The sensitivity to small changes in refractive index for a single-mode 8 μm step-index fiber is quickly appreciated by trying to make suitable adjustments to the sliders for the fiber index and for the cladding index. The 1 to 90 ns pulse width indicates the overlap that can occur due to the variation in path lengths for multi-mode fibers. The scale on the display of the output pulses, shown in Figure 5.5, is automatically adjusted to show the region of interest for input of a single pulse of the specified width.

• **Rainbow**

– **Primary rainbow**
– **Primary and secondary rainbow**

In this option of RAYTRACE the laws of reflection and refraction are applied for rays of class 3 and 4, and the deviation of these rays is plotted against the impact parameter. The change in direction of the ray is plotted from the observer's viewpoint, so that the variation of the angle of the rays to the direction of the incident rays is given, which equates to 180° minus the deviation. In the screen display shown in Figure 5.8 it can be seen that below the graph of impact parameter versus ray angle, there is another graph showing frequency of deviation against ray angle. This second graph effectively shows the intensity of light against ray angle for one color, different colors have maximum intensity at different angles to the incident illumination due to the variation of refractive index with wavelength.

The arc of the rainbow is produced by the constant angle of deviation for each color rotated around the axis of the direction of the incident rays. It will be noticed that no rays of class 3 and 4 are scattered in the range between 42° and 50°, and so this region of sky will appear darker than other regions and is known as Alexander's dark band. There are of course rays of higher class than 3 or 4, but the intensity of these rays will be substantially less, and consequently they do not contribute significantly.

The colors of the rainbow are due to dispersion (the variation of refractive index with wavelength). The spectrum of colors is not pure due to the overlap of the maximum intensity of the different colors. In the program RAYTRACE the dispersion is exaggerated for the sake of clarity. The correct value of refractive index is used for the violet light, and the changes for the other colors are increased by a factor of ten. The standard values of refractive index for each color are given in the Table 5.1.[10] The primary rainbow is produced by rays entering the upper half of the drop, and the secondary rainbow by rays entering the lower half of the drop. The effect of dispersion and the angular spread of the sunlight is to spread the maxima for the different colors over about $2°15'$. The angular width of Alexander's dark band is reduced by a similar amount.

- **Lenses**
  The rays from a point object are traced through the refracting surfaces of the lens system to show the aberrations that occur. If the data file ACHRO-MAT.LEN is available in the chosen directory it will be loaded when the **Lenses** option is selected for the first time. If the file ACHROMAT.LEN is not available, data for a thin meniscus lens will be used. If the data for another lens design is required from a data file, the option **Load lens** in the **File** pull-down menu should be used. Lens design data files should have the extesion ".LEN," and designs can be modified and stored with the same extension. The ray tracing does not take the effects of dispersion into account. The only aberrations shown are the monochromatic aberrations.
  The screen display in Figure 5.13 is divided into three main displays. At the top the rays are shown emerging from the lens, and they will usually be seen to converge to an image point and then diverge. The rays are shown for a distance of one and half times times the focal length. The display gives the value of the focal length and the back focal distance, which is the distance from the plane of the last surface of the lens to the focal plane. There is a slider below this top display which can be used to move the position of a viewing screen along the axis. The position of the viewing screen is indicated by a line across the rays emerging from the lens. The display at the bottom right shows the intersection of the rays with the viewing screen in a spot diagram. The display at the bottom left gives a detailed view of the lens currently being investigated.
  The hot key **F5-Zoom-in Zoom-out** controls the size of the area of the image plane viewed in the spot diagram. **F5-Zoom-in** gives an enlarged view of the image plane and a smaller range on the slider. The center of the rays being viewed may not be in the center of the spot diagram. The hot key **F3-Center** re-centers the spot diagram so that the ray passing through the center of the lens is near the center of the display. This hot key, **F3-Center**, also redraws the slider and adjusts the range so that adjustments can be made to the position of the image screen. The slider is redrawn so that the current position is in the middle of the range. The hot key **F3-Center** can be used to redraw the slider when it is at an extreme of its range and further adjustment is needed.

Table 5.1   Refractive index of water at various wavelengths.

Wavelength nm	Refractive index
480.0	1.3374
508.6	1.3360
546.1	1.3345
589.3	1.3330
643.8	1.3314

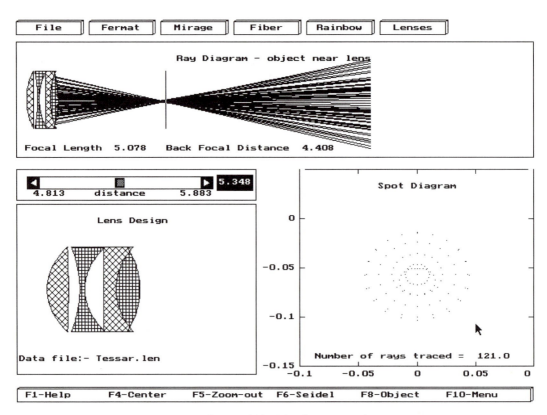

Figure 5.13: The **Lenses** option.

The Seidel aberrations can be displayed for the currently selected lens design by use of the hot key **F6-Seidel**. The spot diagram is obtained again by using the hot key **F3-Center**, **F5-Zoom**, or **F8-Object**. The hot key **F-8 Object** initiates an input screen on which the object position can be set. The three options in the pull-dowm menu are considered in more detail below.

-   **Object at infinity.** In this option 121 parallel rays are incident on the lens in six concentric rings around an axial ray. This is equivalent to a set of rays from an object at infinity. Simple paraxial treatment of the lens leads to the conclusion that a single point image will be formed in the focal plane. It will be seen with this option that when all the rays through the lens are considered there is no longer a unique point of focus and that rays passing through the lens at different radii come to a focus at different points. The rays that are incident on the lens at different radii are shown in the spot diagram in different colors for identification. It will be found that as the plane of the spot diagram is moved through a small range of positions around the position of the paraxial focal point there is a position when the spot diagram approaches closest to a point image; this is the position of the circle of least confusion. The size of the circle of the spot diagram at the paraxial focal point is called the transverse (or lateral) spherical aberration (TSA). The angle which the parallel rays from infinity make with the axis can be set on an input screen which is selected by the hot key **F8-Object**. Thus, it is possible to examine the formation of the off-axis image points and observe the effects of the aberrations, coma, and astigmatism. The effects of curvature of field and distortion will not be apparent for a single point object, although their effect is included in the finite ray trace.

– **Object near lens.** The screen shown in Figure 5.13 is of the **Lenses** option display, which is shown for an off-axis point object near the lens. The ray paths are shown in the top display. The spot diagram shows the intersection of the rays with the screen. The user can adjust the slider to move the screen along the axis of the lens. The position of the point object can be set by the user from an input screen selected by the hot key **F8-Object**. Again the aberrations produced by the lens can be examined by moving the slider which controls the position of the spot diagram.

– **Edit lens design.** On selecting this option the user is presented with an input screen which contains the details of the design of the lens currently being simulated. The number of refracting surfaces and aperture stops is specified at the beginning of the of the input screen. The first lens surface is placed at the origin of the Cartesian axes. The first column of data to be specified is the curvature of the refracting surface. The curvature is positive if the center is to the right of the surface and negative if the center is to the left of the surface. The second column specifies the axial distance to the next refracting surface (or aperture) and should always be positive. The third column specifies the refractive index to the right of the surface specified in the first column. The refractive index should be in the range $1 \leq n \leq 2$. The fourth column need not be used unless the user wants to include stops in the system. A stop is treated as a surface and should be included in the number of refracting surfaces and stops at the top of the input screen. A non-zero value of radius in the fourth column is taken to indicate a stop. The position of the stop is set by the previous line of the input screen. A zero value in the fourth column is taken to indicate a stop of radius 1, which is the standard radius used for the refracting surfaces.

5.14 Running TWOLENS

TWOLENS is a program that can be used to demonstrate simple ray diagrams for a single thin lens, two thin lenses, or spherical mirrors. It is particularly suited for an introductory lecture and demonstration physics courses. In the screen display in Figure 5.11 the object and image are shown, along with the three principal rays that proceed from the object toward the observer. You can use the mouse to move the object, to position the lens or mirror, or to change the focal length of the lens or mirror. The principal rays are redrawn when any of these adjustments are made. The simulation handles converging and diverging lenses and also concave and convex mirrors. Thus, students can quickly get an intuitive feel for real and virtual image formation under a variety of circumstances. The two lenses options show the principal rays for various lens combinations and for a microsope and a telescope. In the case of two lenses the three principal rays for each lens can be drawn to make a total of five rays. The hot key **F2-Full/Scrn** will change the ray diagram so that it fills the display. In this mode the hot key becomes **F2-Split/Scrn** so that the ray diagram is again displayed with a data display below if **F2-Split/Scrn** is pressed. The hot key **F3-Center** will re-center the displayed lens system. Use of the hot key **F4-Zoom-in** produces a rectangle that is controlled by the mouse to select an area for closer examination. The hot key **F5-Zoom-out** is available so that a wider view of the system can be obtained.

- Main menu
  The main menu has six upper-level choices: **File**, **LensType**, **RayPaths**, **Zoom**, **FineTune**, and **Set Data**. The first of these choices provides brief information about the CUPS project and about TWOLENS itself as well as the standard **Configuration** and **Exit Program** selections. The second menu choice, **LensType**, provides access to the lens and mirror ray diagrams and also to the microscope and telescope ray diagrams. The menu item **RayPaths** enables the selection of alternate ray plots. For two lens systems the principal rays of the first lens can be augmented by two extra rays so that the principal rays of the second lens are displayed. This is called a five ray option. For a mirror there is a choice of the third ray plotted: it is either the ray through the center of curvature or it is the ray to the vertex of the mirror which is reflected as though it had met a plane reflecting surface perpendicular to the optic axis.
  The fourth menu option, **Zoom**, provides a range of choices for controlling the size of the screen display. **Manual** leads to an input screen on which the zoom is set by screen coordinates. There are also a number of preset zoom choices for zooming in or out and a **Variable** zoom-in which allows the user to select the zoom with a mouse-controlled rectangle. There are also facilities to pan the image to display different areas of the ray trace. The **Fine Tune** option enables the user to set chosen parameters with sliders that are displayed on the screen. This menu option changes to **Drag Mode** when sliders are displayed and the sub-menu then provides a choice to **Return to drag mode**. The final menu option **Set Data** provides an input screen on which the various parameters may be set. This option could be used in the absence of a mouse, or to check numerical calculations as the resultant image position and size are displayed in the information screen below the ray diagram.

- Mouse control.
  You can use the mouse directly to change the size or position of the object. To do this, just point the mouse close to the tip of the object arrow and click the mouse button. If you hold the mouse button down, you can drag the object continuously to wherever you would like it to be. As the object is moved, the image and the principal rays from the object toward the observer are also continuously adjusted. You can use the mouse to move the lens or the focal point of the lens rather than the object in a similar way, the closest item to the mouse is selected. If you move the focal point to the other side of the lens, it will change from a converging to a diverging lens or vice-versa. The mirror diagram is similar; the main difference is that the rays striking the mirror are reflected rather than transmitted.

The nature of the lens does not change when the object is moved across it to the other side. This is the expected behavior of a physical lens. For convenience and consistency, this simulation treats a mirror in the same way. Thus, when the object is moved to the other side of a concave mirror, it remains concave to the object, rather than becoming convex.

5.15 Exercises

5.1 **Fermat's Principle—Angles of Incidence**
Check that the simulation data displayed is self-consistent by calculating the angle of incidence from the slider data and hence determining $n \sin(i)$ and compare it with the value displayed. The angle of refraction can only be calculated from the slider data when the path taken is at a minimum value. Check the optical path calculation by using Pythagoras' theorem and the value of $n' \sin(i')$ at the minimum path position.

5.2 **Total Internal Reflection**
Within the slider ranges provided by the program, what is the minimum angle of incidence at which total internal reflection will occur?

5.3 **Minimum Path**
Given that the lowest value of the minimum optical path is $\sqrt{2}$, what will the largest value of the minimum optical path be for the range of slider values provided in the program?

5.4 **Preventing Total Internal Reflection**
How would it have been possible to prevent the conditions for total internal reflection arising in the use of the program?

5.5 **Mirage Conditions**
The program illustrates the ray paths for rays near a hot surface for a range of different functions that describe the variation in refractive index with height. Which function is most likely to be of the correct form? Explain the reasons for your choice.

5.6 **Cladding Index**
Determine the value of cladding index which is necessary to ensure single mode propagation in an 8 μm diameter optical communication fiber if the refractive index of the core is 1.500.

5.7 **Use of Multi-Mode Fiber**
By use of the program, estimate the range of distances over which it would be appropriate to use a multi-mode fiber with less than ten modes of propagation for pulses of 90 ns duration with a maximum pulse rate of 1 MHz. Explain the criteria that you have used to make your estimate. Why might it be preferable to use multi-mode fibers in some circumstances?

5.8 **Graphical Method**
No detailed exercises are given for the TWOLENS simulation, because its intended purpose is mainly to demonstrate image formation for a wide

variety of conditions rather than for formal calculations. The simulation can easily be used to demonstrate image formation by convex or concave lenses or mirrors when the object is further away or closer than the focal length. Since the principal rays and the image move in real time as the position of the object or lens or the focal length of the lens is varied, the simulation yields qualitative insight into the effect of these variations. It is perhaps of interest to demonstrate certain specific object/image configurations, such as the symmetric one that brings a real image closest to the object. Finally, the numerical values of the object and image distances and the focal length which are shown on the screen allow one to check that the thin lens equation is satisfied.

5.9   Reducing Spherical Aberration

The amount of spherical aberration when aperture and focal length are fixed varies with lens shape. If the spherical refracting surface is thought of as consisting of a large (infinite) number of prisms arranged around the axis of the lens, it can be seen that the deviation of a ray depends on the orientation of the prism to the incident ray. The incident ray will undergo minimum deviation when it makes the same angle to the surface normal as the emerging ray does to the surface normal. Examine the spherical aberration for a plano-convex lens with the spherical side away from the focal point, and with it toward the focal point. Try to get a focal length of about 5 cm.

References

1. Seidel, L. Astronomische Nachrichten **43**:1856.

2. Meyer-Arendt, J.R. *Introduction to Classical and Modern Optics.* Englewood Cliffs, NJ: Prentice Hall, 1972.

3. Guenther, R. *Modern Optics.* New York: John Wiley and Sons, 1990.

4. Wilson, J., Hawkes, J.F.B. *Optoelectronics—An Introduction.* New York: Prentice Hall, 1989.

5. Tipler, P.A. *Physics for Scientists and Engineers.* 3rd ed. New York: Worth Publishers, 1991.

6. Nussbaum, A., Phillips, R.A. *Contemporary Optics for Scientists and Engineers.* Englewood Cliffs, NJ: Prentice Hall, 1976.

7. Nussbaum, A. *Geometric Optics: An Introduction.* Reading, MA: Addison-Wesley, 1968.

8. Welford, W.T. *Aberrations of Optical Systems.* Bristol: Adam Hilger, 1986.

9. Berry, M., Howls, C. Infinity Interpreted Physics World **6:** 35–39, 1993.

10. Kaye, G.W.C., Laby, T.H. *Tables of Physical and Chemical Constants and Some Mathematical Functions*, 14th ed. London: Longman, 1978.

11. Nussenzveig, H.M. The Theory of the Rainbow. *Scientific American* **236:**116–127, 1977.

12. Longhurst, R.S. *Geometrical and Physical Optics*. London: Longmans Green and Co., 1957.

6

Applications of Interference and Diffraction

Robin A. Giles

> Bodies act upon Light at a distance, and by their action bend its Rays.
> > Query 1: *The Third Book of Opticks*
> > Sir Isaac Newton, 1704

> Rays differing in refrangibility differ in flexibility.
> > Query 2: *The Third Book of Opticks*
> > Sir Isaac Newton, 1704

6.1   Introduction

The program DIFFRACT looks at the basic interference and diffraction effects which underpin the wave nature of light. SPECTRUM allows you to study some of the more practical applications. Four important measuring instruments, the diffraction grating, the prism spectrometer, the Michelson interferometer, and the Fabry-Perot interferometer, are investigated. In addition, the subject of resolving power is considered in relation to these instruments.

6.2   The Diffraction Grating

The diffraction grating consists of a regular array of apertures. Two forms are common, the transmission grating, where the apertures are transparent, and the reflection grating, where the apertures are reflecting.

6.2.1   The Transmission Grating

The Grating Equation

For a plane wave incident on a grating that consists of a large number, N, of identical parallel slits of width w and separation s, the intensity function is given by the equation[1]

$$I_\theta = I_0 \frac{\sin^2 \beta}{\beta^2} \cdot \frac{\sin^2 N\gamma}{\gamma^2}, \tag{6.1}$$

where $\beta = \pi w \sin \theta / \lambda, \gamma = \pi s \sin \theta / \lambda$, N is the number of slits, λ is the wavelength, and θ is the angle with respect to the incident wave.

The second squared term in the equation determines the location of the maxima in the spectrum as a function of the slit separation, whereas the first determines the relative intensities of the lines as a function of the slit width.

Principal maxima occur within the slow variation of intensity due to diffraction from a single slit when $\gamma = m\pi$, $m = 0, 1, 2, \ldots$, i.e., $m\lambda = s \sin \theta$, which is the grating equation, and where m is the order of the principal maximum.

Resolving Power

An important property of a grating is its ability to distinguish between two closely separated wavelengths. The resolving power, R, is given by the equation[2]

$$R = \frac{\lambda}{\Delta \lambda} = Nm. \tag{6.2}$$

Thus, the resolving power increases linearly with the number of slits and the order.

The ability of the diffraction grating to distinguish two wavelengths is limited by overlapping orders. If two wavelengths λ and $\lambda + \Delta \lambda$ have successive orders that are coincident, then

$$(m + 1)\lambda = m(\lambda + \Delta \lambda).$$

The wavelength difference for which this occurs is defined as the minimum *free spectral range* of the grating, $\Delta \lambda_{SR}$, where

$$\Delta \lambda_{SR} = \frac{\lambda}{m}. \tag{6.3}$$

6.2.2   The Reflection Grating

The reflection grating is more commonly used than the transmission type. In this case the grating has a reflecting surface where the transmission grating has an aperture. The latter is opaque, whereas the former is either transmitting or absorbing. The grating equation is unchanged.[2]

$$m\lambda = s \sin \theta. \tag{6.4}$$

In use it is quite common to find that the incident light is not normal to the grating surface, and if θ_0 is the angle of incidence, the modified grating equation becomes

$$m\lambda = s(\sin \theta_0 - \sin \theta). \tag{6.5}$$

As with the transmission grating, the intensity of the individual spectral lines is modulated by the diffraction from a single reflecting strip. This creates a problem in that the energy at higher orders falls off with order and equal amounts of energy go into positive and negative orders. The efficiency of such a grating is greatly improved by *blazing* the grating. Instead of simply having a series of reflecting strips separated by opaque sections or transmitting apertures, a blazed grating is made up of an array of reflecting facets, each inclined the same amount to the plane of the grating (see Fig. 6.1). Initially let us assume that the incident light is normal to the reflecting facet of the grating as indicated in the Figure 6.1. Then, if the angle of inclination of the facets is θ_b—the blaze angle—and the step height, $l \sin \theta_b$ is an integral number of half-wavelengths, then

$$l \sin \theta_b = \frac{p\lambda}{2}, \tag{6.6}$$

where $p = 1, 2, \ldots$ and light reflected off one facet will be in phase with light reflected off the next. Incident light then behaves as if it were reflected back on itself. No energy is diffracted into other orders. The grating equation in this special case is

$$m\lambda = 2l \sin \theta_b, \tag{6.7}$$

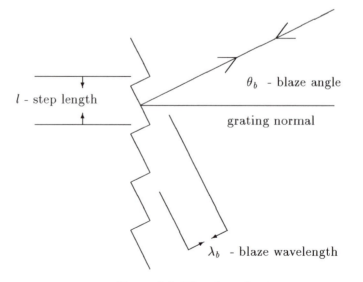

Figure 6.1: Blazed grating.

i.e., compared with the previous equation, $m = p$. The value of λ that satisfies this equation with $p = 1$ is called the *blaze wavelength*, λ_b, and is given by

$$\lambda_b = 2l \sin \theta_b, \tag{6.8}$$

and the blaze wavelength is twice the step height. By choosing the blaze angle appropriately, one particular order in the spectrum can be enhanced and *all* other orders suppressed because whereas the maximum of the intensity envelope created by the blazing lies on the chosen order, the minima of the same function lie on the other orders. If an angle of incidence other than normal is chosen this still applies; the required order is selected but all other orders are suppressed.

6.3   The Prism

The prism is very commonly used in spectral analysis.[5] The right triangular prism is the most common configuration. The angle between the plane boundaries is the *refracting* or *prism angle*. Light entering the first face is refracted according to the relative refractive indices of the media on either side of the boundary. In the case of a glass prism in air (an optically dense medium in an optically rare medium), the incident light is bent towards the normal according to Snell's law. At the second boundary another change of direction occurs.

From the geometry of the path of the light through the prism, the angle of deviation D of the light from the incident direction is a function of the prism angle A and the refractive index of the prism material n and they are related according to the equation

$$n \sin \frac{A}{2} = \sin \frac{(D + A)}{2}. \tag{6.9}$$

Since the refractive index varies with wavelength (dispersion), the angle of minimum deviation will vary with the wavelength of the incident light. Hence, if a wide beam of light enters and contains energy at different wavelengths, the emergent beams will have slightly different directions. However, these beams will overlap and the spectrum will be indistinct. The visibility of the spectrum can be improved by placing a narrow vertical slit in front of the prism and the eye then observes a virtual image of the slit for each wavelength. If the incident light consists of several different discrete wavelengths, a line spectrum is seen in which each line is an image of the slit. In practice, the beams are collimated so that parallel light enters the prism and emerges from it. If a second lens is placed after the prism to form a real image of the slit, then one image for each wavelength is created, producing a spectrum in the focal plane of this second lens which can then be viewed with an eyepiece or displayed on a screen.

The variation of refractive index with wavelength can be expressed over a range of wavelengths for many materials by a simple formula, e.g., the Cauchy formula:

$$n = A + \frac{B}{\lambda^2}. \tag{6.10}$$

Table 6.1   Cauchy constants for commonly used materials

Refractive index data		
Material	A	B
Crown	1.504039	4538.637
Barium flint	1.550231	6209.335
Dense barium flint	1.599661	6251.998
Dense flint	1.625529	10418.74

The Cauchy A and B constants for a set of commonly used materials are listed in Table 6.1.

The ability to resolve two closely separated wavelengths depends on three things, the natural width of the spectral line, the width of the collimator slit, and the size of the prism. The prism resolution program looks at the influence of prism size.[2] The spectral line is assumed to be simple and the slit width small. The influence of these two parameters is left as an advanced exercise.

Consider the case where the spectrometer slit is replaced by a pinhole. If the light was perfectly monochromatic the light would be deviated through a well-defined angle and in the focal plane of the telescope lens you would observe the Fraunhofer diffraction image of the rectangular prism as projected onto a plane normal to the final direction of propagation. With a narrow slit replacing the pinhole there is a superposition of many pinhole images displaced along the direction of the slit. The result is similar to that obtained with a single slit where the slit width is the projected width of the prism, shown as W in Figure 6.2. W is a function of the base length b and the resolving power $R = \lambda/\Delta\lambda$, where λ is the mean wavelength and $\Delta\lambda$ is the wavelength difference that can be resolved, and can be expressed as

$$\frac{\lambda}{\Delta\lambda} = b\frac{dn}{d\lambda}. \qquad (6.11)$$

The dispersive power[3] Δ is a measure of the separation between spectral lines as a function of refractive index and is defined as

$$\Delta = \frac{n_F - n_C}{n_D - 1}, \qquad (6.12)$$

where n_F, n_D, and n_C are the refractive indices for the material for the three Fraunhofer lines, F—486.1 nm (blue line in the hydrogen spectrum), D—589.3 nm (average of the yellow lines in the sodium spectrum), and C—656.3 nm (red line in the hydrogen spectrum).

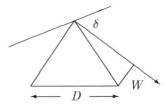

Figure 6.2: Angle of deviation, D, and projected height, W.

6.4   *The Michelson Interferometer*

In the Michelson interferometer,[1] light from an extended source is split into two beams at a partially reflective plate—the beam splitter (Fig. 6.3). One beam travels towards a fixed mirror and is reflected back to the beam splitter. The other beam travels to a moveable mirror, is reflected and also travels back to the beam splitter. The two beams recombine and are then detected, either by eye or by another suitable detector. Looking at the beam splitter from the detector an image of one mirror, M_1, and the actual mirror, M_2, form a layer of thickness d.

Unless d is zero there will always be a path difference between the two beams at the detector creating a phase difference between the combining beams, and interference will result. For the case where d is zero there will, in fact, be a phase difference present because of a phase change on reflection at the beam splitter. Let us assume that the partial silvering is at the first face encountered by the incident beam. Then the beam traveling to mirror M_1 will, on its return to the beam splitter, reflect from a boundary between an optically dense medium (glass) and an optically rare medium (air). As a result there will be a phase change of $\pi/2$ on reflection. When the two beams are detected this phase change of $\pi/2$ will cause there to be no intensity in the field of view. For an effective plate separation other than zero but equal to a multiple of $\lambda/2$, there will be a dark fringe at the center of a pattern of bright circular fringes.

Thus, for an effective plate separation d, the phase difference δ between the combining beams will be given by

$$\delta = \frac{2\pi}{\lambda}(2d)\cos\theta. \tag{6.13}$$

The factor $2d$ results from the fact that the beam travels out and back on reflection from mirror M_2. The maximum order m_{max} occurs at the center of the fringe

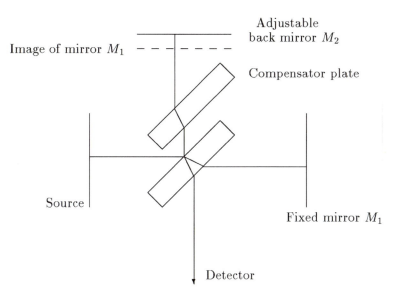

Figure 6.3: Michelson interferometer.

system where $\theta = 0$ and

$$m_{max} = \frac{2d}{\lambda} \qquad (6.14)$$

and the order of the fringes decreases as you move out from the center, which is also the case with the Fabry-Perot interferometer. Note that the diagram includes a compensator plate between the beam splitter and mirror M_2. This plate is exactly the same thickness and is made of the same material as the beam splitter. It is included to make the optical paths, equal to the refractive index of the medium times the physical path length, in the two arms equal. If you follow the paths in the two arms, the beam reflecting off mirror M_1 passes through the beam splitter three times, whereas the beam reflecting off mirror M_2 only passes through the beam splitter once because it is front surface partially reflecting. The compensating plate is necessary if a white light source is used. It is not required if a monochromatic source is used.

The fringe pattern results from the simple superposition of two sinusoidal waveforms, and as a result the intensity of the fringes vary as $cos^2\theta$; i.e., they are much less well defined than is the case for Fabry-Perot fringes. If the incident light contains more than one wavelength, the resulting fringe patterns will combine at the detector. As a result, the visibility of the fringe pattern will change as the effective mirror separation changes. This has been exploited in measuring the difference in wavelength between different components in an incident beam.

In one application, the Fourier transform[6] method can be applied to allow the measurement of wavelengths in a source when more than one or two are present. This technique has proved particularly useful in working in the infrared region of the spectrum. A detector is placed at the center of the field of view. As one mirror is moved relative to the other, the intensity changes according to the varying optical paths in the two arms. The way in which the intensity will vary at the center will depend on the mix of wavelengths in the incident light and their individual intensities. Each wavelength will contribute its own fringe pattern to the overall intensity in the detector. The various sets of $cos^2\theta$ fringes will combine so that the total intensity will vary in a complicated fashion as a function of the effective mirror separation. The output from the detector can be analyzed using the fast Fourier transform[7] (FFT) technique. As a result, it is possible to determine the individual wavelengths that were present in the incident light.

Another application of the interferometer allows the refractive index of a gas to be measured. If, for example, a cell is placed in one arm of the instrument and this cell is evacuated, a fringe system will result in the usual fashion. If gas is now allowed to enter the cell, the optical path length in that arm will increase, and as a result the phase relationship between the two beams at the detector will change. As the optical path in the gas cell changes by $\lambda/2$, one new fringe will be created in the field of view. By counting the number of fringes that pass a given point as a function of pressure, the refractive index of the gas can be calculated. If the incident light contains a large range of frequencies, the visibility of the resulting fringe pattern will be very poor; and indeed it is likely that the overlapping fringe patterns will give an essentially uniform intensity in the field of view at the detector. There is one exception when the effective plate separation is zero. In this case there will be a black field of view at the center surrounded by a blurred set of fringes.

Because the gas cell must necessarily have entrance and exit windows made of a transparent material there will be dispersion effects if more than one wavelength is used.

The Michelson interferometer can also be adapted to allow for the measurement of the refractive index of a solid in the form of a parallel-sided sample. Just as there is a change in the optical path length when gas was allowed to fill the gas cell, there will be a similar change in optical path length if the sample is introduced into one arm, but unlike the gas technique, where the change in optical path length is gradual as the pressure changes, the change in optical path length this time is sudden. However, we can use the fact that if the optical path lengths in the two arms are equal, the field of view is completely black. The method requires that the two paths be made equal with no sample present. The sample is then placed in one arm, making the optical path length for a beam traveling in that arm larger. A set of fringes will be seen. The mirror in the other arm is then moved back just enough to make the optical paths equal again and the black field reappears. In this application it is usual to tilt one mirror slightly with respect to the other which has the effect of changing the fringe pattern from circular to straight (localized fringes), and a straight black central fringe is bounded by straight colored fringes.

6.5   *The Fabry-Perot Interferometer*

The Fabry-Perot interferometer[5] employs multiple reflections to create interference. It can be used to measure wavelengths with high precision and it also allows for the study of fine structure of atomic spectra. Two optically flat, partially reflecting plates (commonly of glass or quartz) are separated by a small gap and held accurately parallel. If the plates have a fixed separation, the unit is called an *etalon*. If the separation distance can be varied the device is called an *interferometer*. A very high order of flatness is required if maximum fringe sharpness is to be obtained. A truth of the order of up to 1/100 of a wavelength is necessary. The device is mounted between a collimating lens and a focusing lens. If an extended source of light is used, interference fringes in the form of concentric circles result. The condition for the production of an interference maximum is

$$2t \cos \theta = m\lambda, \tag{6.15}$$

where θ is the angle subtended by the fringe and t is the etalon gap. If the plates are partially reflecting, the fringes are very sharply defined and high resolution can result. The intensity varies with the angle θ according to the equation

$$I_\theta = \frac{I_{max}}{\left(1 + F \sin^2 \frac{\theta}{2}\right)}. \tag{6.16}$$

The factor $F = 4r^2/(1 - r^2)^2$, where r^2 is the reflectivity and F is called the *contrast* and it allows the relative transmission to be expressed as the Airy function, I/I_{max}. The peak in the transmission occurs when the thickness t is equal to a multiple of $\lambda/2$. The sharpness of the transmission curve increases as r increases and for values of r near unity the fringes are exceptionally sharp. The accuracy with which the interferometer can measure a wavelength is called the *chromatic resolving power*, $\mathcal{R}$, and is defined as $\lambda/\Delta\lambda$, where

$$\mathcal{R} = \frac{\lambda}{\Delta\lambda} = \frac{m\pi}{2}\sqrt{F}. \tag{6.17}$$

From Eq. 6.15, the maximum value of the order occurs at the center of the fringe pattern and

$$m_{max} = \frac{2t}{\lambda}. \tag{6.18}$$

If the thickness of the gap is changed, the fringe pattern changes. If the gap increases, the diameter of the fringes increases as the fringes move out from the center. The peaks in the fringe pattern occur when the spacing is a multiple of $\lambda/2$. Every time t changes by $\lambda/2$ another fringe is created at the center.

The *free spectral range* $\Delta\lambda_{SR}$ is defined as the wavelength separation between adjacent orders of interference and

$$\Delta\lambda_{SR} = \frac{\lambda}{m} \tag{6.19}$$

which is the same result as obtained with the diffraction grating Eq. 6.3.

Increasing the separation between the plates increases the resolving power but with this increase comes a decrease in the free spectral range. If two wavelengths are separated by a difference greater than the free spectral range, an incorrect value for their separation will result because adjacent fringes will not be of the same order. Thus, increased resolution comes at the expense of being able to determine wavelength uniquely. An auxiliary filter is usually introduced to limit the range of wavelengths entering the instrument. The ratio of the *free spectral range* to the minimum resolvable wavelength difference is called the *finesse* $\mathcal{F}$ and is given by the expression

$$\mathcal{F} = \frac{\Delta\lambda_{SR}}{\Delta\lambda} = \frac{\pi}{2}\sqrt{F}. \tag{6.20}$$

6.6   Running SPECTRUM

6.6.1   General Comments

On startup you will see an introductory screen. Pressing **Enter** or clicking the mouse will clear this information screen and display the Gratings: Transmission Grating-Spectrum program. You can continue with this or select an alternative program by choosing from the top **Menu**.

The first menu column **File** consists of five options. **About CUPS** provides general information on the CUPS project. The second option, **About Program**, provides a general introduction and credits. The third option, **Program Details**, provides some general information about the programs available to you. The fourth option, **Configuration**, controls system parameters and is fully discussed in chapter 1. The final option, **Exit Program**, allows you to leave the program. The various sections of the programs can be accessed using the other **Menu** options.

On the bottom of the screen is a set of hot keys. There are the always available **F1-Help** and **F10-Menu** hot keys. Each program has its own set of hot keys for special purposes as appropriate. **F1-Help** provides context-sensitive help appropriate to the particular program you have chosen. **F10-Menu** allows you to activate the top line **Menu**. In addition you will normally find a **F9-Defaults** hot key, which runs the program with the original default values. Sometimes values are changed so much that it is often quicker to reset the defaults and then start again. Some programs have other hot keys, which are discussed in the relevant sections.

On selecting a **Menu** item, one of two things will happen. In a few cases (in **Prism** and **Michelson**) the first thing you will see is an **Input Screen**. This will allow you to enter parameters for that program. The input screen has data windows with default values which can be changed. The simulation then runs on clicking on the **OK** button or pressing **ENTER**. In this case an extra hot key **F8-Rerun**, is provided that will allow you to return to the input screen with the previously selected values. The **F9-Defaults** hot key returns the input screen with the original default values. If there is no input screen you will see the program for a set of default parameters. New parameters may then be set using the sliders. In the majority of cases, the simulation runs on release of the slider thumb. A greater insight into many of the simulations can often be obtained by pressing the mouse button while the mouse arrow is located on either end of the slider. In this case, the variable is smoothly changed and the plots change accordingly.

6.6.2   Menu Item: Gratings

Transmission Grating—Spectrum

Default Settings:
Normal incidence
Slit Separation—0.003 mm
Slit Width—0.001 mm
First Wavelength—650 nm
Second Wavelength—500 nm

The viewport displays the diffraction spectrum. The slit width, the slit separation, and the two wavelengths can be changed using the sliders. If the mouse is moved into the viewport you will be able to measure diffraction angles and intensities (normalized to one at 0°).

Transmission Grating—Resolution

Default Settings:
Normal incidence
Number of Slits—2
Slit Width—0.005 mm
Slit Separation—0.02 mm
First Wavelength—600 nm
Second Wavelength—550 nm

The upper viewport shows the intensity variation as a function of angle for the two wavelengths separately (in false color) and the lower viewport shows them combined (in white). The number of slits, the slit width, the slit separation, and the two wavelengths can all be changed from this screen using the sliders. On the right-hand side there are two buttons and on the bottom of the screen is a set of four hot keys. The buttons **Zoom Range** and **Measurement** determine whether you want to expand the angular range of the graphs or wish to make measurements on the total intensity graph in the lower viewport. With the **Zoom Range** button chosen, if the mouse is used to select a new low angle and then a new high angle, in either viewport, the scale will be expanded between these new values. The new angles can be selected in either order. The **F4-Zoom Out** hot key will return the default angular range. To make measurements in the lower viewport, the **Measurement** button must be selected. You will then be able to measure the intensity by moving the mouse into the lower viewport.

Blazed Reflection Grating

> **Default Settings:**
> Angle of Incidence—30°
> Blaze Angle—8.50°
> Wavelength—550 nm
> Steplength—4000 nm

The upper viewport shows the unmodified diffraction pattern for the default settings. In addition there is a white envelope curve which shows the effect of blazing the grating. The lower viewport shows the same spectrum as modified by the blazing envelope. The selectivity that results is evident. Sliders allow you to change the incident angle, the blaze angle, the wavelength, and the steplength. The incident angle and the blaze angle sliders each have a zoom capability to allow finer adjustment. The mouse can be used in the lower viewport to allow for the measurement of the angles of diffraction. The blaze wavelength is also displayed. Note that when the order 1 is selected, the blaze wavelength is equal to the wavelength of the incident light, for order 2 it is double, etc.

6.6.3   Menu Item: Prism

Spectral Response

> **Default Settings:**
> Prism Angle—60°
> Incident Angle—60°
> Prism Material—Crown glass
> Source—Helium

The input screen allows you to set the prism angle and the incident angle. A set of buttons allows you to choose from one of four prism materials and one of six light sources. If you wish you can select the **Choose Material** button to enter your own choice of prism material using a new input screen into which you enter the appropriate Cauchy A and B constants, or a different light source selecting the **Choose Source** buttons and using another new input screen that requires you to enter up to eight wavelengths. If fewer than eight wavelengths are defined, the number should be entered on the input screen, and any extra values that might be in the lower part of the list on the input screen will be ignored on acceptance.

By selecting one of the listed prism materials, the Cauchy A and B constants for the default prism material are entered as preselected-Eq. 6.10. A list of these constants can be found in Table 6.1. The default wavelength sets[4] are shown in Table 6.2. On selecting **OK** the program will run with the chosen prism material and light source. The upper viewport shows the spectrum for the chosen set of wavelengths and prism material. The angular range can be varied using the **Zoom Range** button. If selected, the complete set of spectral lines are displayed in an appropriately adjusted range. Figure 6.4 shows a typical screen where the **Zoom Range** button has been used to allow the spectral lines to be more clearly seen. You can return to the default range by pressing the **Normal** button. The angle of incidence can be varied by a slider and the sensitivity can be changed for fine adjustment using the **Normal** and **Zoom** buttons. You might find it interesting to use the mouse to change the angle of incidence in a smooth, continuous fashion. If all of the lines are totally internally reflected in the high resolution mode and you select a new angle of incidence, you will be returned to low resolution mode. You can then reselect the high resolution mode when there are spectral lines to be seen. In the lower right-hand viewport is a ray diagram for the prism. Adjustment of the incident angle shows as a prism rotation and new ray paths are drawn. In the event that total internal reflection occurs, this information is shown in the upper viewport and also in the ray diagram. The angle of deviation of a particular line can be measured by moving the mouse into the viewport.

A table containing angular information can be displayed if the **F5-Data** hot key is used. This data should be read in conjunction with the ray diagram in Figure 6.5, where the angles β, γ, and δ are defined. The other notation in the figure will be helpful if you look at the Pascal code. The data table is automatically removed if a new angle of incidence is chosen.

Resolution

Default Settings:
Prism Angle—60°
Incident Angle—60°
Base Length—0.05 m
Prism Material—Crown glass
Source—Sodium

Table 6.2   Wavelengths for Helium, Hydrogen, Mercury, Sodium, Krypton, and Lithium (nm)

Wavelength data					
Helium	Hydrogen	Mercury	Sodium	Krypton	Lithium
667.8	656.6	579.1	615.7	769.4	670.8
587.5	486.1	577.0	589.5	768.5	610.4
501.5	434.0	546.2	588.9	760.1	460.3
492.2	410.1	491.7	568.6	758.7	413.3
471.3	397.0	436.0	498.0	587.0	398.5
447.3	388.9	407.9	466.6	557.0	391.5
438.9	383.5	404.8	454.3	450.2	
402.7	379.7		442.1	445.3	

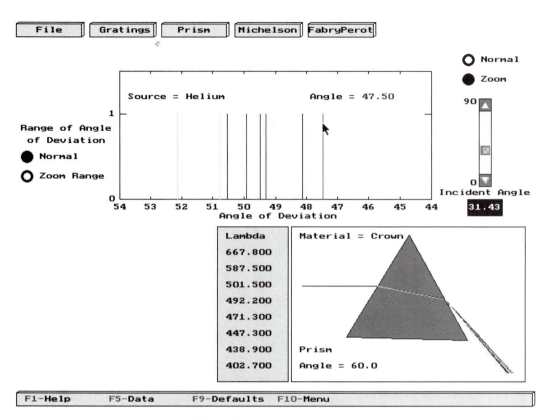

Figure 6.4: Prism: Spectrum.

The input screen allows you to choose the prism angle, the incident angle, and the prism base length. One of four prism materials can be selected together with one of four light sources that have strong doublets. They are the krypton red doublet, the sodium yellow D lines, the mercury yellow doublet, and the potassium violet doublet. The default wavelengths[4] are listed in Table 6.3. The Cauchy constants are those used in the **Prism-Spectrum** program and are to be found in Table 6.1. As in that program you can also enter the parameters for a different prism material and/or a new source by selecting the **Choose Material** and **Choose Source** buttons and entering the parameters on the appropriate input screen. The upper viewport shows the spectrum for each wavelength separately in false color in an appropriate angular range. The middle viewport shows the combined intensity. The incident angle can be changed using a slider whose sensitivity can be changed using the **Zoom** and **Normal** buttons. A slider is also provided that will allow you to change the baselength of the prism without the need to return to the input screen. A data table is provided in the bottom viewport.

Table 6.3   Wavelengths for Krypton, Sodium, Mercury, and Potassium doublets (nm)

Wavelength data			
Krypton	Sodium	Mercury	Potassium
769.454	589.592	579.066	404.721
768.524	588.995	576.960	404.414

6.6.4 Menu Item: Michelson

Spectrum

Default Settings:
Initial Effective Mirror Separation—0.001 m
First Wavelength—600 nm
Second Wavelength—650 nm
Mirror Travel Distance—6e-6 m
Angular Range——$-5°$ to $+5°$ (fixed)

The input screen allows you to choose the initial effective mirror separation and two wavelengths. For an animation option, the total distance that the rear mirror is to travel is chosen. There are two buttons on the input screen. The first, **Spectrum**, is set to true. The other button should be selected for the exercises. On selecting the input screen, the upper viewport shows the intensity across the field of view for both wavelengths in false color. The lower viewport shows the combined intensities in white. With the two default wavelengths you will readily recognize a beat pattern. The slider allows you to change the effective mirror separation. By clicking the mouse on a fringe in the upper viewport you will be able to measure the order of that fringe. Note that the order of a fringe at a maximum is half-integral. The **F3-ShowRings** hot key will allow you to see the ring structure in the lower viewport. The rings can be cleared and the original fringe pattern returned using the **F4-ClearRings** hot key.

A typical screen is shown in Figure 6.6, where the **F3-ShowRings** hot key has been used to display the ring system for two different wavelengths. You will notice how the fringe visibility varies across the field of view due to overlap.

An animation mode can be selected using the **Animation ON/Animation OFF** buttons. When a suitable choice of parameters has been established and the **Animation ON** button is selected, a new screen will appear which simulates the appearance of the fringe pattern as the rear mirror is moved steadily backwards from the initial position over the selected range of mirror travel. The appearance of the fringe pattern is shown at regular intervals, the position being indicated in the viewport. For this program an expanded set of hot keys is provided. The usual **F1-Help** and **F10-Menu** hot keys are now accompanied by three additional keys—**F2-Run** or **F2-Stop**, **F4-Faster** and **F5-Slower**. Use of these hot keys will

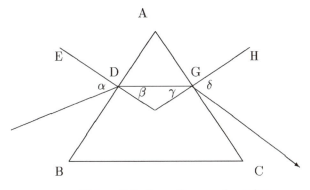

Figure 6.5: Ray diagram for prism.

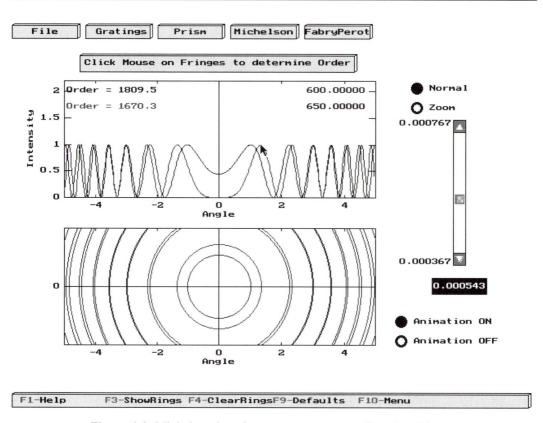

Figure 6.6: Michelson interferometer—spectrum: Showing Rings.

allow you to freeze the motion and adjust the speed of the mirror travel, useful if you are working with a too-slow or too-fast computer. However, be careful, because if you speed up the motion too much you will see a false result and the fringe pattern can appear to move in the wrong direction (like the wheels of the stage-coach in Western movies). The current position of the mirror is shown in the viewport.

Fourier Transform Spectroscopy

> **Default Settings:**
> First Wavelength—1200 nm
> Second Wavelength—2400 nm
> Mirror Travel Distance—3.6e-5 m
> Number of Transform Points—256

This program provides a simple introduction to the principles of Fourier transform spectroscopy. On selecting this option, the opening screen shows two sliders that can be used to choose two wavelengths in the incident light and a third slider that controls the distance travelled by the moveable mirror. A set of buttons provides a way of changing the number of Fourier transform sampling points over the distance that the mirror travels. There are two viewports. The upper viewport shows the intensity variation at a detector located at the center of the field of view as the travelling mirror moves over the predetermined distance. The lower viewport the result of applying a Fourier transform to the intensity waveform. The results are shown for the default settings. A pair of **Welch Window Buttons** will be seen in the upper viewport. There are two additional hot keys **F2-Problem d** and **F3-Problem e** for exercise use.

The intensity variation displayed is clearly the superposition of two wavetrains of different wavelengths. In order to see how the method works, set both wavelengths to the same value (4000 nm), make the mirror travel distance a simple and small multiple of the wavelength(s) (0.0000400 m), and turn the Welch window **OFF**. The use of the Welch window is discussed below. The result will be the sum of two identical cos^2 intensity variations over the range of mirror travel and, because you have chosen simple parameter values, an exact number of wavelengths will fit into the mirror travel distance. Since $cos^2\theta$ equals $(1 + \cos 2\theta)/2$ the waveform will have an average value of 1.0 and its Fourier series will have both a constant term and sinusoidal components.

The bottom viewport shows the frequency components present in the sampled intensity variation. The horizontal axis is calibrated as a frequency in terms of the number of complete waveforms that could be fitted into distance that the mirror moved. A constant term is found in location 0 representing the finite average value of the intensity pattern. Thus, you should find that, for the simple case under analysis, there will be a sharp peak at a low value and also at the zero value. By moving the mouse into the lower viewport, the wavelength of incident light will be displayed. The relationship between the frequency and the wavelength is a simple function of the mirror travel distance.

Now run the default case again. Here one wavelength is double the other and the mirror travel distance is a simple multiple of both wavelengths. You will again readily recognize the intensity variation versus mirror distance as the superposition of the two incident wavelengths. You will now see three peaks in the lower viewport, one at zero and two others. The wavelengths relative to the latter two peaks can be measured using the mouse as before. Note that the height of these peaks is the same because of an inherent assumption in the program that the two wavelengths were of equal intensity.

You will probably notice that the wavelength calculation seems somewhat inaccurate. This is a result of the sensitivity of the value of the calculated wavelength to the horizontal position of the mouse when the mirror travel distance is small. Increase the mirror travel distance but still keep it a simple multiple of the larger of the two selected wavelengths. The intensity variation will probably now look more complex, although all that has happened is that more cycles of the waveform are displayed. You will now find that the results show at higher frequencies because more complete periods are displayed in the intensity variation. However, you will see an improvement in the accuracy in the wavelength calculation because of the higher precision now available using the mouse.

Now, of course, in the real world the incident light would normally consist of wavelengths which do not bear a simple relationship and the mirror travel distance cannot be chosen to be a simple function of these wavelengths. As a result, whereas before the periodicity of the intensity function was such that integral number of waveforms fitted into the mirror travel distance, this will almost certainly no longer be the case. Hence, the initial value of the intensity function will no longer be the same value as the final value, and there is a discontinuity or step in the function at its ends. You can study this effect using the Fourier analysis and Fourier transforms programs included in this part of the CUPS project, using the **Draw Curve** mode and dragging the mouse across the viewport and deliberately ending the trace at a different level from the level that you started at. In the Fourier series analysis of a function with a step, other frequency components are required to try to accommodate the discontinuity. With the Fourier transform of a curve with a similar discontinuity, rather than a sharply peaked frequency distribution which you get with a smooth function, with a step you get a broadening.

It is common to use a special window the force the end points to be equal. The Welch[7] window uses the function:

$$w_j = 1 - \left(\frac{j - (N - 1)/2}{(N + 1)/2} \right)^2, \tag{6.21}$$

where j is the sample number and N is the total number of sampling points. In the example case here there is an extra consideration. The Welch window function assumes that the curve to be modified is approximately symmetrically disposed about the horizontal axis. In our case the curve oscillates about a line 1.0 units above the horizontal axis. To overcome this in using the window, the data is manipulated to remove the 1.0 term, then the Welch window is applied and then the 1.0 term is added back. The result is clear on the screen. The beginning point is equal to the end point at 1.0. You can switch the Welch window on and off using the **Welch Window** button. You should immediately see the effect on the intensity variation and the change in the results of the FFT procedure.

There is another problem to face - *aliasing.*[7] If you choose too large a mirror travel distance in relation to the wavelengths in the incident light you will begin to find that the wavelength calculations are again incorrect but for a different reason—aliasing. In order for the FFT procedure to work without aliasing it is necessary for there to be at least two sampling points per cycle of the waveform for each wavelength in the incident light. If there are fewer than two sampling points per cycle an incorrect result will occur. If there are fewer than two sampling points per cycle of the intensity variation function, the frequency components calculated using the Fourier transform procedure will be incorrect and the wavelengths are incorrect as a result. You can study this using the simulation. In the case of two wavelengths, the shorter wavelength is inaccurately determined before the longer wavelength.

As a lead into studying this phenomenon, set both of the wavelengths at 4000 nm, leave the mirror travel distance at the default value 3.6e-5 m and the number of sampling points at the default value 256. The number of complete waveforms inside the upper viewport is 18, and this is the number calculated by the Fourier transform. Check the wavelength using the mouse and the correct result is shown. Change the number of sampling points both up and down. Check the number of waveforms in the upper viewport and the frequency and wavelength in the lower viewport. Everything checks out. Now increase the mirror travel distance to 1e-4 m for the same wavelengths and 256 sampling points. Check the number of waveforms, the frequency, and the calculated wavelength. Repeat for the other numbers of sampling points. You will find that correct results are obtained for all but the lowest number of sampling points—64. Now look at the waveform that has been analyzed as a function of the number of sampling points. The true waveform is displayed for the four highest values, although something funny seems to be happening with 128 sampling points. With 64 points the sampled intensity variation is clearly not the real one. By sampling at too infrequent an interval the variation with distance is modified. The Fourier transform gives the correct number of periods for the intensity variation displayed but aliasing has created an inappropriate intensity variation and hence the wavelength calculation is incorrect. The exercises include some questions addressing this important issue.

Refractive Index of a Gas

Default Settings:
Wavelength—600 nm
Cell Length—0.01 m
Gas—Air
Angular Range—$-2°$ to $+2°$ (fixed)

The input screen is used to select the interferometer parameters. You can choose the wavelength of the monochromatic incident light and the length of the gas cell. You can also select the gas that will go into the cell. For purposes of the program there is a default gas—namely, air—and we will assume that we know the refractive index[4] normalised to STP (1.0002926). Of course, in a real experiment, the refractive index of the gas would be the unknown. The exercise set will ask you to select the buttons labeled **Measure Gas A** and **Measure Cell Length**.

On acceptance the viewport shows the fringe pattern in the field of view as a function of the pressure as it increases from zero to approximately one atmosphere. As in **Spectrum** in the animation mode, the usual **F1-Help** and **F10-Menu** hot keys are now accompanied by three additional keys—**F2-Run** or **F2-Stop**, **F4-Faster** and **F5-Slower**. Use of these hot keys will allow you to freeze the motion and adjust the rate of change in pressure. The final pressure is determined by the rate that the program runs but will be not less than 760 mm Hg. The default rate speed setting will give a final pressure of exactly 760 mm Hg (101.325 Kpascal). The viewport also contains information on the order of the initial and current fringes relative to the center of the field of view. The refractive index value can then be calculated from the cell length and the mirror travel distance. It is conventional to quote an index relative to one standard atmosphere. The parameter values and the gas in the cell are indicated on the screen.

Refractive Index of a Solid

Default Settings:
Back Mirror Position—0.0015 m
First Wavelength—550 nm
Second Wavelength—600 nm
Sample Thickness—0.001 m
Angular Range—$-5°$ to $+5°$

The input screen allows you to choose the initial effective mirror separation, two wavelengths, and the thickness of the solid sample. The default material is crown glass. The value of the refractive index used is that averaged over the wavelength range (600–650 nm). The other button, **SOLID A**, is for use in the exercises.

On acceptance the screen resembles the one you will have seen in the first section—**Spectrum**. The upper viewport shows the appearance in the field of view, and the lower viewport the intensities combined. A slider is provided so that you can change the position of the back mirror to allow for a change in the optical path lengths. This slider can be used in both a low sensitivity mode **Normal** and a high sensitivity mode **Zoom** using the upper set of buttons, as in other programs. Another pair of buttons at the bottom right-hand side **Sample In** and **Sample Out** lets you introduce or remove the sample into one arm.

The default value places the mirror such that the two path lengths are not equal and a fringe pattern results. By moving one mirror using the slider you can equalize the two paths and obtain the black field. If the sample is then introduced, the fringe pattern reflects the sudden change in optical path length in one arm. Adjusting the position of the back mirror will put the black field back in view. You will probably need to use the high sensitivity setting for the final adjustment. If an even finer setting is required you can type in the desired value in the slider viewport and press the **Enter** key. Since the added path length P is simply

$$P = (n - 1)t, \qquad\qquad (6.22)$$

where t = the sample thickness, it is a simple matter to calculate the refractive index if the sample thickness is known.

Initially use monochromatic light by setting both wavelengths to the same value but later try two different wavelengths. With two wavelengths you will find that the setting of the moveable rear mirror that was correct for monochromatic light is also correct for the two wavelengths and, we can conclude, for white light. With the optical paths equal in both arms, all wavelengths create the black field.

6.6.5 Menu Item: Fabry-Perot

Spectrum and Resolution

> **Default Settings:**
> Reflectivity—0.9
> Plate Separation—0.001 mm
> First Wavelength—500 nm
> Wavelength Increment—0.17 nm

The upper viewport shows the individual fringe intensities in false color. The lower viewport shows the combined fringe intensity in white. New parameter values can be selected using the sliders which give control of the reflectivity, the plate separation, one wavelength (an integer value), and the difference betweeen this one wavelength and a second. There is a set of five hot keys at the foot of the screen which are used in conjunction with two buttons—**Zoom Range** and **Measurement**. With the **Zoom Range** button chosen, the mouse can be used to select a new low angle and then a new high angle in *either* viewport, the angular range is expanded between the new values. Selecting **F4-Zoom Out** returns the original range. This expansion can be repeated if the initial set of low and high angles is not appropriate. The same facility is also used in the **Transmission Grating-Resolution** program. If the **Measurement** button is selected, the mouse can be clicked on a fringe in the upper viewport to determine the order of that fringe. If used in the lower viewport, you can use the mouse to measure the intensity at any point on the curve. This will allow the Rayleigh resolution criterion to be tested. The **F3-ShowRings** hot key will allow you to see the ring structure in the lower viewport. The original fringe pattern can be returned by using the **F4-Zoom Out** hot key.

6.7 Exercises

6.7.1 Transmission Grating: Spectrum

a.   Run the program to see the effect of changing the slit width, the slit separation, and the wavelength.

b.   Under certain conditions the spectral lines for the different orders start to overlap. Using the default settings for slit width, slit separation, and wavelength, what is the smallest separation of two wavelengths that could be used so that there is no overlap as far as the third order?

c.   A grating was used in third order with incident light from a helium discharge tube. The red line (667.8 nm) in third order diffracts at an angle of 42°. How many lines per millimeter did the grating have?

d.   A spectral line due to red light of wavelength 650 nm in the second order lies exactly on a second line due to green light of wavelength 510 nm in the third order. What was the slit separation of the grating?

e.   A diffraction grating is used to study the spectrum from a source containing two discrete wavelengths. Choose suitable values for slit width and slit separation to confirm the relationship between intensity of the lines in the various orders as a function of the width of the individual slits.

6.7.2 Transmission Grating: Resolution

a.   Run the program to see the way in which the resolution of the grating depends on the slit width, the slit separation, the number of slits, and the order.

b.   It is necessary to separate the two sodium D lines (588.995 nm and 589.592 nm). Using the Rayleigh criterion that the diffraction maximum of one line should lie on the first minimum of an adjacent line, find suitable grating parameters such that the required resolution is available in the first order. Repeat for the second and third order. Confirm that the ratio of the intensity minimum between two lines to the maximum for lines that are just separated according to the Rayleigh criterion is 0.81. Show that your results are consistent with the definition of resolving power (Eq. 6.2).

c.   What is the smallest separation between two wavelengths that can be resolved around 550 nm in the third order using a transmission grating that is 2.5 cm wide and is ruled with 50 lines per millimeter?

6.7.3 Blazed Reflection Grating

a.   Run the program to study:

   i. the relation between wavelength and selectivity

    ii. the relation between step length and selectivity
    iii. the effect of changing the blaze angle
    iv. the effect of changing the angle of incidence.

b.   A light source has two component wavelengths—500 nm and 700 nm. Design a suitable reflection grating that will be able to select the longer wavelength and completely suppress the shorter wavelength. Will it operate for all orders? Will it operate for all incident angles?

c.   What blaze angle should a grating ruled with 750 lines/mm have to select the first order of diffraction at 650 nm?

6.7.4   Spectrum

a.   Run the program to study the effect of changing the angle of incidence over the complete range from 0° to 90°. Note how the angular separation of the deviated rays across a spectrum varies with the angle of incidence and the material of the prism. Study the effect of varying the prism angle.

b.   Confirm Eq. 6.9, the minimum deviation, formula for a series of prism angles and prism materials.

c.   Make a series of measurements on the spectrum produced by one of the available sources using one of the available prism materials. Use this data to calculate the Cauchy A and B constants for that material by plotting a suitable graph. Repeat for another source and a second prism material.

d.   What is the smallest angle of incidence that could be used for the longer of the two sodium D line wavelengths so that it is just not internally reflected using a dense flint prism? Confirm by a direct ray tracing calculation.

e.   Calculate the dispersive power of a crown glass prism and a dense flint prism using Eq. 6.12. Can you establish any link between the calculated values and the angular separation of the C and F lines at minimum deviation?

f.   Figure 6.7 shows an arrangement of three prisms called the *Amici system* that is the basis of the direct vision spectroscope. In one such design the first and third prisms are made of the same material but are different from the center prism. A beam of light entering from the left emerges in the same direction but with dispersion. Design such and instrument using two crown glass prisms (cr) and one of dense flint (df).

6.7.5   Resolution

a.   Run the program to study how the appearance of two spectral lines varies as a function of prism base length using the Rayleigh criterion. As an example, take the sodium D lines (588.995 nm and 589.592 nm).

b.   A prism with a base length of 2.5 cm is made of barium flint glass. Use Eq. 6.11 to calculate the resolving power at 600 nm. Confirm your results directly by using the program.

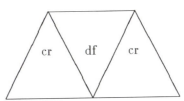

Figure 6.7: The Amici prism system.

c.   What is the smallest prism that could be used to resolve the two
     yellow lines in the mercury spectrum for an angle of incidence of 45°.
     The prism is made of dense barium flint and the prism angle is 30°.
     Repeat for a 30° prism made of crown glass. Compare your results
     and show that your results are consistent with Eq. 6.11.

6.7.6 Michelson Interferometer: Spectrum

a.   Run the program to study how the fringe structure produced by the
     interferometer depends on the plate separation and wavelength. Use
     the animation mode to simulate the motion of the back mirror.
b.   Use the interferometer to measure the difference in wavelength
     between the two sodium D lines.
c.   In a remarkable experiment carried out by Meissner,[8] the rate at
     which a plant grew was investigated using a Michelson interferometer.
     One mirror was attached to the top of the plant and the moveable
     mirror was adjusted to give a black field of view indicating that the
     effective plate separation was zero. As the plant grew the effective
     plate separation changed so that fringes appeared. At regular intervals
     as the plant grew the position of the moveable mirror was changed to
     return the black field and from the distance that the mirror had to be
     moved the rate of plant growth was calculated. Select the **Problem**
     button from the input screen and accept the data. Set the lower button
     to **Start of Growth** and find the position of the moveable mirror
     to give the black field. Then select the **Finish of Growth** button,
     simulating a time interval of 8 hours during which the plant grew
     and its mirror moved. Find the distance that you have to move the
     other mirror to return the black field and hence calculate the rate of
     plant growth in metre/sec and mm/hour. Take the wavelength of the
     monochromatic source to be 550 nm.

6.7.7 Fourier Transform Spectroscopy

a.   Run the program as suggested in the text, pages 186–189.
b.   Make a thorough study of the phenomenon of aliasing. For a given
     pair of wavelengths, find the maximum distance that the mirror can
     travel before aliasing becomes a problem. Show that the result is
     consistent with the need for there to be at least two points per cycle
     of the shortest waveform present in the source.

c.   Study the results as a function of the number of sampling points and the mirror travel distance in the absence of aliasing. In particular study the manner in which the Welch window works when the beginning and end points of the intensity variation are not the same. Why is it that when the start and finish values are the same the Welch window is of no assistance?

d.   Run the simulation and use the **F2-Problem d** hot key. An unknown light source has been used with the interferometer. Analyse the result, making sure that you have used suitable values of the mirror travel distance and number of sampling points. Measure the wavelengths in the source. Can you identify the source?

e.   Use the **F3-Problem e** hot key and identify the light source used in a manner similar to d) above.

f.   Study the effect of having a waveform which starts and finishes at a different level using the **Draw Curve** option in the FOURIER program.

6.7.8   Refractive Index of a Gas

a.   Run the program to see how the fringes produced in the interferometer vary as a function of the parameters.

b.   Select **Gas A** from the input screen and measure its refractive index using a gas cell 0.015 m in length. Identify the gas.

c.   A gas cell was filled with hydrogen. Select **Measure Cell Length** from the input screen and determine the length of the gas cell, if the refractive index of the gas was 1.000132.

6.7.9   Refractive Index of a Solid

a.   Run the program to establish the relation between the parameters of the interferometer.

b.   For this problem select **Solid A** from the input screen. With the sample button set at **Sample Out** there is no sample in one of the arms of the interferometer. First find the position of the moveable mirror to give the black field. Then place the sample into arm with the fixed mirror by setting the sample button at **Sample In**. Find the new position of the moveable mirror to return the black field. If the thickness of the sample was 0.001 m, what was its refractive index? Can you identify the solid?

6.7.10   Fabry-Perot Interferometer Spectrum and Resolution

a.   Run the program to study how the fringe structure produced by the interferometer depends on the plate separation and wavelength.

b.   What is the least separation between two wavelengths (in the region 600 nm) that could be resolved using an etalon that is 5 mm thick with a reflectance of 85%?

    c.   An etalon has a reflectance of 90%. What plate separation would be needed to separate the sodium D lines?

    d.   An etalon has a reflectance of 85% and a thickness of 2 mm. It is used with a source that contains just two wavelengths, 550.0 nm and 549.5 nm. In which order would you have to use the etalon to separate the two lines?

    e.   Investigate the variation of the finesse F of an etalon with the reflectivity r^2.

    f.   Investigate the free spectral range, $\Delta \lambda_{SR}$, of an etalon.

    g.   Investigate the variation of the resolving power, $\mathcal{R}$, with the contrast F.

6.8   Further Developments Involving Optional Programming

If you are going to attempt the advanced programming exercises below, make sure that you have made copies of the original code. Keep these originals safe and only develop the copies. Do not be deterred if a new program does not run or it crashes. You might be wise to study the CUPS Utilities Manual.

    a.   Modify the Transmission Grating program to allow for angles of incidence other than 0°.

    b.   Consider the effect of a finite slit width on the spectrum produced by a prism. How would the resolution be affected? If you are able to write Pascal code, make the necessary modification to the code to show this effect.

    c.   In the prism resolution program an idealised source is used, one with no line width. The next approximation would be a line with a Gaussian shape factor. What impact would that have on the resolving power of the prism? If you are able, make the neccesary modification to the Pascal code to show this effect also.

    d.   If you have completed the previous question, what about two lines close in wavelength with Gaussian shape factors?

    e.   In the Michelson interferometer, mention was made of localized fringes which are straight and not circular. Modify the Pascal code to show localized fringes.

References

1. Jenkins, F.A., White, H.E. *Fundamentals of Optics.* New York: McGraw-Hill, 1976.

2. Klein, M.V. *Optics.* New York: John Wiley and Sons, 1970.

3. Born, M., Wolf, E. *Principles of Optics.* London: Pergamon Press, 1980.

4. Hodgson, C.D., Weast, R.C., Selby, S.M. *Handbook of Chemistry and Physics,* 37th ed. Cleveland: Chemical Rubber Publishing Company, 1955.

5. Longhurst, R.S. *Geometrical and Physical Optics.* New York: John Wiley and Sons, 1964.

6. Steward, E.G. *Fourier Optics.* Chichester: Ellis Horwood, 1983.

7. Press, W.H., Flannery, B.P., Teukolsky, S.A., Vetterling, W.T. *Numerical Recipes in Pascal.* Cambridge: Cambridge University Press, 1989.

8. Meissner, K.W. Interferometer for the Investigation of the Process of Growth, *Physikalische Zeitschrift* **30**:965, 1929.

7

Electromagnetic Waves

Ronald Stoner

"It takes a much higher degree of imagination to understand the electric and magnetic fields than to understand invisible angels... I speak of the E- and B-fields and wave my arms, and you may imagine that I see them... (But,) I cannot really make a picture that is anything like the true waves."

Richard Feynman in *The Feynman Lectures, Vol. II*

7.1 Introduction

The presence throughout space of the universal microwave background demonstrates that electromagnetic radiation can carry energy and information over truly astronomical reaches of space and time. The persistence of electromagnetic waves far from their source is possible because the electric and magnetic fields reinforce each other in their respective oscillations: Time variation in the electric field is the "displacement current" source of the magnetic field, and time variation of the magnetic field produces the electric field by Faraday induction.

The fields that constitute electromagnetic radiation are difficult to visualize or to portray in the static, two-dimensional format of textbook line drawing, because the precise ways the electric and magnetic fields interact in radiation is inherently three-dimensional and dynamic. Computer simulation and animation provide a way to visualize the dynamic, three-dimensional character of their interaction. This CUPS simulation uses perspective projection and animation to allow the user to visualize various polarization states of coupled harmonic waves of electric and magnetic fields, both in free space and in matter. The simulation also can be used to illustrate the phenomena of reflection and transmission at a boundary, to observe differences between travelling and standing waves, and to investigate the effects of optical elements such as polarizers and quarter-wave plates.

7.2   Electromagnetic Waves in Vacuum

Maxwell's equations govern the behavior of the electric field $\vec{E}$ and the magnetic field $\vec{B}$ in the classical regime of electromagnetism. In vacuum, these equations take the following simple form

$$\vec{\nabla} \cdot \vec{E} = 0, \tag{7.1}$$

$$\vec{\nabla} \times \vec{E} = \frac{-1}{c} \frac{\partial \vec{B}}{\partial t}, \tag{7.2}$$

$$\vec{\nabla} \cdot \vec{B} = 0, \tag{7.3}$$

$$\vec{\nabla} \times \vec{B} = \frac{1}{c} \frac{\partial \vec{E}}{\partial t}. \tag{7.4}$$

The above form of Maxwell's equations is appropriate when cgs (Gaussian) units are used for the fields,[2] and are convenient for treating electromagnetic wave phenomena; the form of Maxwell's equations appropriate for the case of rationalized mks units can be found in standard textbooks.[1] In the cgs system, $\vec{E}$ and $\vec{B}$ have the same units (Gauss), which has particular advantages for the analysis of electromagnetic radiation. Among the particular solutions of the free-space Maxwell's equations are those in which both the electric field, $\vec{E}(\vec{r}, t)$ and the magnetic field $\vec{B}(\vec{r}, t)$ have the mathematical forms of polarized, travelling plane waves, namely,

$$\vec{E}(\vec{r}, t) = \vec{E}_o e^{i(\omega t - \vec{k} \cdot \vec{r})} \tag{7.5}$$

$$\vec{B}(\vec{r}, t) = \vec{B}_o e^{i(\omega t - \vec{k} \cdot \vec{r})}. \tag{7.6}$$

The symbols $\vec{E}_o$ and $\vec{B}_o$ in these plane wave solutions are the amplitudes of a plane harmonic wave, which are vectors with complex components. The use of complex numbers and exponential functions allows the mathematical description of the harmonic waves to be done more compactly. It is implicitly understood that the actual fields are the real parts of the complex functions on the right-hand sides of the Eqs. 7.5 and 7.6. The $\vec{k}$ in the above plane wave formulas is the wave vector ($|\vec{k}| = \frac{2\pi}{\lambda}$), and $\omega = 2\pi f$ is the angular frequency of the wave.

Since Maxwell's equations in matter-free space admit plane wave solutions, they can be thought of as the wave equations for electromagnetic waves in space. Since the equations connect partial space derivatives of one field with partial time derivatives of the other, the plane waves for $\vec{E}(\vec{r}, t)$ and $\vec{B}(\vec{r}, t)$ are not independent. The parameters in the plane wave description of $\vec{E}(\vec{r}, t)$ completely determine the parameters in the $\vec{B}(\vec{r}, t)$ plane wave. The connection can be established by substitution of the traveling plane wave forms into Maxwell's equations, which reveals that they represent possible solutions only under the following conditions:

$$\vec{k} \cdot \vec{E}_o = 0 \tag{7.7}$$

$$\vec{k} \times \vec{E}_o = \frac{\omega}{c}\vec{B}_o \tag{7.8}$$

$$\vec{k} \cdot \vec{B}_o = 0 \tag{7.9}$$

$$\vec{k} \times \vec{B}_o = -\frac{\omega}{c}\vec{E}_o. \tag{7.10}$$

These mathematical conditions imply that the three vectors $\vec{k}$, $\vec{E}_o$, and $\vec{B}_o$ are mutually perpendicular, and that the magnitudes of $\vec{E}_o$ and $\vec{B}_o$ are equal. They further imply that $\omega = ck$, which means that electromagnetic waves in space are dispersionless, and that the phase velocity of the waves is $\lambda f = c$, the speed of light.

It is always possible to choose a frame of reference in which $\vec{k}$ is in the $+z$-direction, in which case $\vec{E}_o$ and $\vec{B}_o$ have only x- and y-components. The restrictions that the magnitudes of these two vector amplitudes are equal and that the two vectors are mutually perpendicular still leaves considerable latitude, since each has two complex components. It is the relative amplitudes and phases of the two Cartesian components of $\vec{E}_o$ that determine the polarization of the wave. The conventions and symbols used below to represent this polarization are those used in the standard advanced text by Born and Wolf.[3] It is always possible to choose the origin and the zero of time in such a way that the real parts of the two components of $\vec{E}$ can be written

$$E_x = a_1 cos(\omega t - \vec{k} \cdot \vec{r}) \tag{7.11}$$

and

$$E_y = a_2 cos(\omega t - \vec{k} \cdot \vec{r} + \delta), \tag{7.12}$$

where a_1 are a_2 are real amplitudes, and where δ is the difference in phase between the x- and y-components of the electric field. In this case, the complex values of the two components of $\vec{E}_o$ and $\vec{B}_o$ are

$$E_{ox} = a_1, \tag{7.13}$$

$$E_{oy} = a_2 e^{i\delta}, \tag{7.14}$$

$$B_{ox} = a_2 e^{i\delta}, \tag{7.15}$$

and

$$B_{oy} = -a_1. \tag{7.16}$$

The parameter δ represents the radian measure of the difference in phase of oscillation of the two components of $\vec{E}$ (or of $\vec{B}$). If $\delta = 0$, then the two components oscillate in phase and the wave is plane polarized. If $\delta = \frac{\pi}{2}$, then the y-component of $\vec{E}$ comes to a maximum one-fourth of a period before the x-component, and the radiation is said to be right-circularly polarized (i.e., negative helicity). If $\delta = -\frac{\pi}{2}$, then the x-component of $\vec{E}$ comes to a maximum one-fourth of a period before the y-component, and the radiation is said to be left-circularly polarized (positive helicity). If $\delta = \pi$ (or $-\pi$), the wave is again plane polarized, but in a direction perpendicular to the $\delta = 0$ case. Intermediate values of δ correspond to various degrees of "elliptical" polarization.

7.3   Stokes' Parameters

Maxwell's equations and the particular choices for the origins of space and time leave three degrees of freedom for describing a monochromatic electromagnetic plane wave. These three degrees of freedom correspond to the freedom to choose values of the three parameters a_1, a_2, and δ. In the case of fully polarized, coherent electromagnetic radiation, these three parameters can be related to measurable quantities that are traditional for characterizing the polarization state of electromagnetic radiation. These quantities are the Stokes parameters,[3,4] which can be related to a_1, a_2, and δ as follows:

$$s_0 = a_1^2 + a_2^2, \tag{7.17}$$

$$s_1 = a_1^2 - a_2^2, \tag{7.18}$$

$$s_2 = 2a_1 a_2 \cos(\delta), \tag{7.19}$$

$$s_3 = 2a_1 a_2 \sin(\delta). \tag{7.20}$$

There are traditionally four independent Stokes' parameters measured in the laboratory to characterize the polarization state of electromagnetic radiation, but only three of the four parameters defined above are truly independent, since they are derived from only the three free parameters a_1, a_2, and δ. The fourth degree of freedom corresponds to the "degree of polarization" of a "partially polarized" beam of radiation. In the case of partially polarized radiation, the quantities a_1, a_2, and δ are not constant, but functions of time that change very slowly over many, many periods of wave oscillation. But a_1, a_2, and δ may be quite variable over the much longer time scales required to measure the Stokes' parameters in the laboratory. This means that Stokes' parameters measured in the laboratory are actually time averages, at least in the partially polarized case.

For similar reasons, the mathematical description and computer-assisted visualization of partially polarized radiation is beyond the scope of this simulation. This is because partially polarized radiation cannot be truly monochromatic and coherent. While it may be possible to think of a plane wave whose polarization state changes slowly with time, the Fourier spectrum of such a wave would necessarily have a finite frequency width. This frequency width corresponds to the rate at which the polarization state changes. In this sense, partially polarized radiation is only "quasi-monochromatic."

In the laboratory, quasi-monochromatic radiation is produced by noncoherent sources and has a very narrow frequency spectrum. The narrowness of the frequency distribution implies that the radiation is coherent over a very large number of periods of oscillation. It is effectively impossible to simulate such quasi-monochromatic radiation by the computer animation technique employed here, at least with any degree of realism, since to do so would require generating an extremely large number of periods or wavelengths simultaneously. For that reason, the present simulation is confined to the treatment of fully polarized (i.e., exactly monochromatic) radiation, meaning that the quantities a_1, a_2 and δ are constant in time. Therefore the Stokes' parameters are not independent, but subject to the condition that

$$s_0^2 = s_1^2 + s_2^2 + s_3^2. \tag{7.21}$$

Equations 7.17 through 7.20 are compatible with Eq. 7.21, which is the constraint that reduces the degrees of freedom from four to three for "fully polarized" or "purely monochromatic" radiation.

All four Stokes' parameters have the dimension of intensity, i.e., radiative power per unit area. The overall intensity of a plane electromagnetic wave is given by the magnitude of the Poynting vector, $\vec{S} = c\vec{E} \times \vec{B}$, which, averaged over one period of oscillation, is $\frac{c}{2}(a_1^2 + a_2^2)$. It follows that the Stokes' parameter, s_0, is a measure of the overall intensity of the electromagnetic wave. The other three Stokes' parameters have the following meanings:

- s_1 represents the difference between the intensity transmitted by a linear polarizer with polarizing axis in a given direction (say, 0°) and the intensity transmitted when the polarizer is rotated to a perpendicular direction (90°).

- s_2 is similar to s_1 except that the mutually perpendicular directions of the polarizer axis are 45° and 135°.

- s_3 represents the difference between the intensity of the radiation after passing through a right-circular polarizer and its intensity after passing through a left-circular polarizer.

7.4   Electromagnetic Waves in Matter

When matter is subjected to electric and magnetic fields, electric and magnetic dipoles are induced in the individual atoms. The electric and magnetic polarization of the atoms represents microscopic electric charges and electric currents that themselves are sources of electric and magnetic fields. Over distance scales much larger than atomic sizes, these induced dipoles appear as continuous electric polarization and magnetization fields in the material, and the microscopic charges and currents appear as macroscopic distributions of "bound" electric charge and current. Further, an electric field applied to a material with mobile microscopic charge carriers, such as the free electrons in metals, will generate "free" electric currents that can be described by a electric current density $\vec{J}(\vec{r}, t)$.

The effect of the applied fields in most materials is "linear," which means that the polarization fields and free currents are proportional to the fields that induce them. Since all these induced charges and currents are both produced by the fields and are themselves sources of the fields, they are coupled with the electromagnetic wave. They oscillate coherently with the electric and magnetic fields, they carry part of the energy and momentum, and so in a real sense are part of the electromagnetic wave in the material.

It is often mathematically convenient in the presence of matter to introduce two new macroscopic fields, $\vec{D}(\vec{r}, t)$ and $\vec{H}(\vec{r}, t)$, and to use them to rewrite Maxwell's equations in such a way that bound charges and bound currents do not appear as explicit sources of $\vec{E}(\vec{r}, t)$ and $\vec{B}(\vec{r}, t)$. This version of Maxwell's equations, in the absence of free electric charge, is[1,2]

$$\vec{\nabla} \cdot \vec{D} = 0, \tag{7.22}$$

$$\vec{\nabla} \times \vec{E} = \frac{-1}{c} \frac{\partial \vec{B}}{\partial t}, \tag{7.23}$$

$$\vec{\nabla} \cdot \vec{B} = 0, \tag{7.24}$$

$$\vec{\nabla} \times \vec{H} = \frac{1}{c} \frac{\partial \vec{D}}{\partial t} + \frac{4\pi}{c} \vec{J}, \tag{7.25}$$

where $\vec{J}$ is the density of free current.

The figure below (Fig. 7.1) shows a sample screen from section of this simulation that depicts the behavior of electromagnetic waves incident from free space onto an isotropic, homogeneous, linear material. In linear isotropic materials, the atomic polarizations are both parallel to and proportional to the applied $\vec{E}$ and $\vec{B}$ fields, and Ohm's law describes the free electric current produced by an applied electric field. Maxwell's equations can be further simplified in this case by writing

$$\vec{D} = \epsilon \vec{E}, \tag{7.26}$$

$$\vec{B} = \mu \vec{H}, \tag{7.27}$$

and

$$\vec{J} = \sigma \vec{E}, \tag{7.28}$$

where ϵ, μ, and σ are real constants that characterize the material. They have the following meanings:

- ϵ is the relative electric permittivity,
- μ is the relative magnetic permeability, and
- σ is the electrical conductivity.

In cgs units, ϵ and μ are dimensionless pure numbers, identical to the relative permittivity and permeability in rationalized mks units. But $\vec{E}$ and $\vec{J}$ are of different dimension, so Ohm's law (Eq. 7.28) defines the conductivity σ to have the dimension of current density per unit field. This is inconvenient for computation, and it makes the numerical value of σ depend on the system of units chosen. For those reasons, the simulation employs and asks the user to supply a dimensionless σ that is proportional to the real conductivity of the material. More specifically, in cgs units, σ has the dimension of frequency; however, ω is fixed in the simulation. The code defines $\omega = 1$ implicitly for computational efficiency, while the apparent "real time" frequency of oscillation is fixed by computer clock speed and the user-supplied time increment between animation frames. The implicit unit adopted for σ is $\frac{\omega}{4\pi} = f/2$. Therefore, to simulate a particular material and frequency, the user should substitute the dimensionless ratio 2(conductivity/frequency) for the σ in the data input screen.

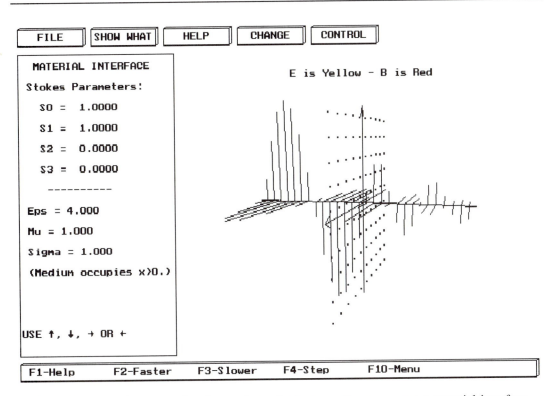

Figure 7.1: Sample screen showing a plane electromagnetic wave near a material interface.

If the material is homogeneous, ϵ, μ, and σ are uniform in space, and have no space derivatives inside the material. In this case, Maxwell's equations can again be written in terms of $\vec{E}$ and $\vec{B}$ as follows:

$$\vec{\nabla} \cdot \vec{E} = 0, \tag{7.29}$$

$$\vec{\nabla} \times \vec{E} = \frac{-1}{c} \frac{\partial \vec{B}}{\partial t}, \tag{7.30}$$

$$\vec{\nabla} \cdot \vec{B} = 0, \tag{7.31}$$

$$\vec{\nabla} \times \vec{B} = \frac{\epsilon\mu}{c} \frac{\partial \vec{E}}{\partial t} + \frac{4\pi\mu\sigma}{c} \vec{E}. \tag{7.32}$$

Implicit in these equations is the assumption that ϵ, μ, and σ are real constants, which is a valid assumption only if the current density and polarization fields respond immediately to changes in the applied fields. In other words, the equations are valid if the time scales for variations of the fields are much longer than both the natural periods of atomic oscillations and the time intervals between scatterings of the free charge carriers in the material.

Equations 7.29 through 7.32 are approximations used in this simulation for the partial differential equations of electromagnetic waves in matter. Substitution of the plane wave forms of Eqs. 7.5 and 8.6 into them produces the following condition on the wave vector k and frequency ω of the plane wave:

$$k^2 = \frac{\omega^2}{c^2}(\mu\epsilon + i\frac{4\pi\mu\sigma}{\omega}), \tag{7.33}$$

where ω is real by virtue of the implicit condition that the fields at a given point in space are simple harmonic functions of time. The imaginary part of the right-hand side of this equation represents the effect of the current density term in Maxwell's equations. The oscillations of the current density are one-quarter period out of phase with the other fields, which provides a mechanism for the dissipation of the energy carried by the wave. The presence of this term means that k will have both real and imaginary parts:

$$k = \beta + i\alpha, \tag{7.34}$$

where both β and α depend on the material properties ϵ, μ, and σ, and on the frequency ω. The imaginary part, α, leads to the exponential attenuation of the wave along the direction of propagation, and the real part, β, is the phase gradient in the material, i.e., $\beta = \frac{2\pi}{\lambda}$.

7.4.1   Reflection and Transmission at a Boundary

Straight lines perpendicular to the axis of propagation, the z-axis, are used throughout this simulation to represent the fields $\vec{E}$ and $\vec{B}$ evaluated at points along that axis. Because the x- and y-dimensions are used this way to represent the magnitudes and directions of the fields, it isn't feasible to use the x- and y-dimensions to convey additional information, such as changes in direction of a plane wave upon reflection or refraction at an interface. On the other hand, reflection and transmission of a wave normally incident on a plane boundary occurs along the axis of propagation, so it can be represented.

When a plane electromagnetic wave is incident on the interface between free space and matter, Maxwell's equations must be satisfied at the interface. This leads to boundary conditions that relate $\vec{E}$ and $\vec{B}$ on opposite sides of the interface. The boundary conditions require, in general, that the incident wave be partially reflected and partially transmitted. For example, the requirement that $\vec{\nabla} \times \vec{E} = \frac{-1}{c}\frac{\partial \vec{B}}{\partial t}$ at the interface, together with the physical requirement that $\frac{\partial \vec{B}}{\partial t}$ be finite, requires for normal incidence that neither $\vec{E}$ nor $\frac{\partial \vec{E}}{\partial t}$ change discontinuously across the boundary. The abrupt change in magnetization at the interface means that there is no similar requirement that $\vec{B}$ be continuous.

Consider the special case of an incident wave with $\vec{k} = k\hat{z}$. This wave will be transmitted across an interface lying in the x-y plane into the medium with the same frequency ω, but with a generally different wave vector $\vec{k}_t = k_t\hat{z}$. In order to match the above boundary conditions on $\vec{E}$, the incident wave must also be partially reflected out the boundary, leading to a reflected plane wave with $\vec{k}_r = -k\hat{z}$. The boundary conditions on the electric field then connect the superposition of incident and reflected wave on one side of the boundary with the transmitted wave on the other.

The code for the simulation of reflection and transmission at a material interface proceeds in the following way: The Stokes' parameters of the incident plane wave determine the parameters a_1, a_2, and δ for it through the inverse of Eqs. 7.17 through 7.20. The parameters ϵ, μ, and σ are then used to find the real and imaginary parts of k_t. Next, the boundary conditions on $\vec{E}$ are used to determine the amplitudes and phases of both the x- and y-components of $\vec{E}$ and $\vec{B}$ of the reflected and transmitted waves. This finally allows $\vec{E}$ and $\vec{B}$ to be computed at all times t and positions z along the z-axis.

7.5 Polarizers and Quarter-Wave Plates

Polarizers are devices which resolve incident electromagnetic radiation into components that are parallel and perpendicular to the optical axis of the polarizer, then transmit only the parallel component. Quarter-wave plates are devices that resolve incident radiation into components parallel and perpendicular to an optical axis, then introduce a relative phase shift of 90° ($\frac{\pi}{2}$ radians) before combining and transmitting them again.[4] This may be accomplished in different ways in different frequency bands, but in the case of visible light, polarizers and quarter-wave plates are many wavelengths thick. In the present simulation these devices are represented by drawings of very thin square plates.

Depending on the orientation of its optical axis, a perfect quarter-wave plate can convert a plane-polarized plane wave to a circularly polarized wave without change in intensity, and vice-versa. Figure 7.2 is a sample screen that depicts this process.

A polarizer can also convert an incident elliptically or circularly polarized wave to a transmitted plane-polarized wave, but this generally involves a decrease in intensity since only one of the two components of the incident wave is transmitted.

The code for this simulation computes the effect of a polarizer in the following way: As in the other cases simulated, the polarization of the incident wave is characterized by Stokes' parameters, from which a_1, a_2, and δ can be computed for the incident wave. These parameters are used to create a two-component vector representing the complex amplitude $\vec{E}_o = (a_1, a_2 e^{i\delta})$. This two-component vector is successively multiplied by three 2×2 matrices representing the following operations: 1) rotation to bring the plane of polarization parallel to the optic axis of the polarizer; 2) removal of the component of $\vec{E}_o$ perpendicular to the optic axis; and, 3) rotation of the plane of polarization back to the original direction. The result is a two-component vector for the complex amplitude of the transmitted wave, $\vec{E}_{ot}$, which is then converted to parameters a_{1t}, a_{2t}, and δ_t for the transmitted wave.

A very similar procedure is used to represent the quarter-wave plate, the only difference being that the second matrix multiplication, instead of removing the y-component, introduces a quarter-wave phase shift by multiplying it by $e^{i\frac{\pi}{2}} = i$. The effects of various combinations of polarizers and wave plates can be treated in much the same way through successive matrix multiplication.

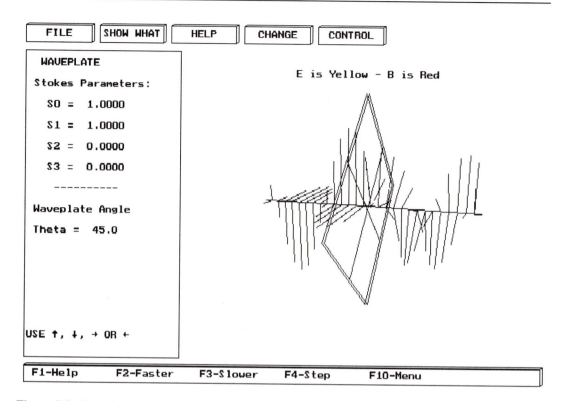

Figure 7.2: Sample screen showing conversion of a plane polarized wave into a circularly polarized wave by a quarter-wave plate.

7.6 Using the Simulation

Each frame of the simulation shows a perspective projection of coordinate axes and color-coded vectors representing $\vec{E}$ and $\vec{B}$ at uniformly spaced intervals along the z-axis. Animation is achieved by frame flipping. While one frame is displayed, another hidden one is being drawn that corresponds to a slightly later time. Once the drawing of the hidden frame is complete, it is displayed, and the previous frame is hidden, erased, and redrawn to show the fields at a still later time.

The perspective projection is drawn with the origin at screen center and from the point of view of an observer a fixed distance from the origin. Pressing the arrow keys moves the observation point in angular increments of 5° in altitude and azimuth relative to the x-z plane. This allows the user to examine the wave from many different points of view so as to see its three-dimensional structure.

The ability to change the user's viewpoint can be used effectively to examine a single component of the fields: Viewing the wave from a point on the x-axis will yield a projection of the x-components of the fields on the screen, and viewing from a point on the y-axis will produce a projection of the y-components. Similarly, the nature of the polarization (plane, circular, or elliptical) may be most obvious by viewing it from a point on either the positive or negative z-axis.

Some of the simulations have a hot key labeled **Rotate**, which allows the user to change the orientation of a polarizer or waveplate in 5° increments "on the fly." This has the pedagogic advantage of allowing "real time" simulation of optics exercises. Each such shift in orientation of a polarizer will break the coherence of the simulated wave, and may result in an instantaneous large jump in phase. This is not unphysical since it would be impossible in practice to rotate a polarizer by any significant amount during a fraction of one oscillation of an optical wave. Similar phase jumps may result when using hot keys labeled **Faster** and **Slower**.

7.7 *Exercises*

7.1 **The Stokes' parameters**

The first Stokes' parameter is fixed at $s_0 = 1$ in this simulation, and the other three are automatically normalized to satisfy $s_0^2 = (s_1^2 + s_2^2 + s_3^2)$. Use the simulation to find and record the values of sets of the parameters (s_1, s_2, and s_3) that produce each of the following types of polarization:

   a. Plane polarization in the x-direction
   b. Plane polarization in the y-direction
   c. Plane polarization at a 45° angle with the x-direction
   d. Right circular polarization
   e. Left circular polarization
   f. Elliptical polarization with semimajor axis double that of the semi-minor axis.

7.2 **Are $\vec{E}$ and $\vec{B}$ in phase?**

Examine the phase relationship between $\vec{E}$ and $\vec{B}$ for a plane polarized wave ($s_3 = 0$). Note that they oscillate in phase. Show that this is required by Maxwell's equations (Eqs. 7.1–7.4) because space derivatives of $\vec{E}$ are proportional to time derivatives of $\vec{B}$ (and vice-versa). Next examine a circularly polarized wave ($s_1 = s_2 = 0$, $s_3 = 1$) from a viewpoint along either the x-axis or the y-axis. Explain why $\vec{E}$ and $\vec{B}$ appear to be 90° out of phase from this perspective, and why this is consistent with Maxwell's equations.

7.3 **The speed of electromagnetic waves in matter**

The speed (phase velocity) of a harmonic wave is the wavelength times the frequency: $v_{phase} = f\lambda = \frac{\omega}{k}$. Use the simulation to measure the wavelength of a plane wave in insulating matter (i.e., $\sigma = 0$) for several values of μ and ϵ. This can be done by choosing a viewpoint far from the origin and in the y-z plane, then measuring distance with a ruler on the computer screen. Given that the frequency in all cases is the same, and given that the speed of radiation in free space is c, determine the speed of the wave in matter in each case. Plot $log(v_{phase})$ vs. $log(\mu)$ for $\epsilon = 1$ and $log(v_{phase})$ vs. $log(\epsilon)$ for $\mu = 1$. From your plot, infer a formula for v_{phase} as a function of μ and ϵ. Given that neither energy nor information can travel faster than c, do you believe it is possible for the speed of an electromagnetic wave in matter to be larger than c?

7.4 **Boundary conditions at an interface**

Satisfy yourself by watching the simulation that $\vec{E}$ varies continuously across the interface between free space and matter for all values of μ, ϵ, and σ. Try values of μ and ϵ in the range 0.5 to 20.0, and values of $\sigma > 0$. Also, in the case $\mu = 1$, $\epsilon = 1$, and $\sigma = 0$, there is no discontinuity, so $\vec{B}$ is also continuous across the boundary in that special case. Can you find any other sets of μ, ϵ, and σ that allow $\vec{B}$ to vary continuously across the interface? Given that $\vec{B} = \mu\vec{H}$, can you determine conditions under which $\vec{H}$ is continuous across the interface?

7.5   Reflection, transmission, and standing waves

Matter-free space in this simulation corresponds to $\mu = 1$, $\epsilon = 1$, and $\sigma = 0$; if the material occupying the region $z > 0$ has different values of these parameters, a plane wave incident on the interface will be partially reflected and partially transmitted into the medium beyond the interface. In the latter case, the superposition of incident and reflected waves in the region $z < 0$ produces a partially standing wave. The amplitude of the reflected wave can be inferred by noting the sizes of the maxima and minima in the standing wave envelope for a plane-polarized incident wave. These maxima and minima are, respectively, the sum and difference of the amplitudes of the incident and reflected waves.

a. Make estimates of the amplitude of the reflected wave for enough values of μ, ϵ, and σ to sketch a plot of the reflection coefficient (defined here to be the ratio of the reflected amplitude to the incident amplitude) as a function of each one of these parameters, in each case leaving the other two at their free-space values.

b. When seismic waves are reflected from interfaces between two layers of rock in geologic studies, the reflection coefficient depends only on the ratio of the speeds of sound on either side of the interface. Make measurements of the reflection coefficient and wavelength in the medium for various values of μ and ϵ to see whether the reflection coefficient for electromagnetic waves at the interface depends only on the speed of the wave inside the medium. If so, make a plot of reflection coefficient vs. speed of the wave in the medium.

c. For $\sigma = 0$, sketch a plot of the transmission coefficient (ratio of the transmitted amplitude to the incident amplitude) as a function of μ and of ϵ.

d. For very large values of μ, ϵ, and σ the incident wave is totally reflected. Compare the wave pattern produced in the region $z < 0$ when the incident wave is plane polarized with the pattern produced when the incident wave is circularly polarized. Make a sketch of the energy density $(E^2 + B^2)/2$ as a function of z for each of these cases.

e. In comparing the circularly-polarized and plane-polarized cases in part d above, you should have noticed that the electric and magnetic fields were essentially parallel in the resulting standing wave in one case, but perpendicular in the other. Explain how the electric and magnetic fields in an electromagnetic wave can ever be parallel, given the statement on page 199 that "the three vectors $\vec{k}$, $\vec{E}_o$ and $\vec{B}_o$ are mutually perpendicular." By looking at the standing wave along the z-axis, estimate the angle between the two fields in standing waves produced by elliptically polarized incident waves, and make a rough plot of that angle vs the Stokes parameter s_3 of the incident wave.

7.6 **Skin depth**

For $\sigma > 0$, the transmitted wave decays exponentially with distance into the region $z > 0$. That is, the amplitude of the oscillations in $\vec{E}$ varies as $e^{\frac{-z}{\delta}}$, where δ is called the "skin depth" or "penetration depth." By varying the values of μ, ϵ, and σ determine the approximate functional dependence of δ on these three quantities. Sketch rough plots of δ vs. each quantity.

7.7 **Reflection by real materials**

Look up the permittivity, permeability, and conductivity of some real materials like glass, semiconductors, and metals. Convert the conductivity to cgs units ($\sec^{-1}$). Also look up the reflectivity of the same materials for radiation at different frequencies. Simulate the reflection and transmission of radiation from these materials to see if the model used to create the simulation is roughly accurate. Try to explain any differences in terms of the validity of the approximations made in Section 7.4.

7.8 **How to measure the Stokes' parameters**

a. Find an orientation of a quarter-wave plate that will convert plane-polarized radiation to right-circularly polarized radiation.

b. Find an orientation of a quarter-wave plate that will convert plane-polarized radiation to left-circularly polarized radiation.

c. Verify that s_1 and s_2 can be measured using a single polarizer arranged at angles 0°, 45°, 90°, and 135°. Remember that intensity is proportional to the square of the amplitude.

d. Construct a scheme for measuring s_3 and use the simulation to verify that it works.

References

1. Lorrain, P., Corson, D.P., Lorrain, F. *Electromagnetic Fields and Waves.* San Francisco: Freeman Press, 1988, chaps. 27, 28.

2. Jackson, J.D. *Classical Electrodynamics.* New York: John Wiley and Sons, 1975, chap. 7.

3. Born, M., Wolf, E. *Principles of Optics*, 4th ed. London: Pergamon Press, 1970, chaps. 1, 10.

4. Guenther, R.D. *Modern Optics.* New York: John Wiley and Sons, 1990, chaps. 2, 13.

8

One-Dimensional Oscillator Chain

Wolfgang Christian, Susan Fischer, Andrew Antonelli

8.1 Introduction

A chain of coupled oscillators is an excellent topic to end the study of wave phenomena, since such a system demonstrates how atoms—albeit Newtonian atoms—influence wave phenomena. This system models a simple crystal lattice and exhibits solid-state effects such as band gaps, as well as many of the wave effects studied in previous chapters, such as traveling waves, normal modes, dispersion, and—with the addition of nonlinear restoring forces solitons.[*] By letting the mass density, $\frac{\Delta m}{\Delta x}$, become a continuous function of position, x, a lattice model (Fig. 8.1) can be a useful starting point for the derivation of various wave equations. Related lattices are not hard to imagine. They occur in many branches of physics: mechanics, masses or pendula coupled by springs; electricity and magnetism, a series of LRC circuits; atomic physics, atoms in a laser cavity; acoustics, volumes connected by tubes. Since masses and springs are easy to visualize, we develop the equations of motion for this system, but a simple change of variables is all that is needed to adapt our system to these other problems.[4]

Let the motion of each mass, M_j, be restricted to a one-dimensional transverse displacement, $y_j(t)$, about its equilibrium point. Each mass is then coupled to a neighboring mass through a force, $F_{interaction}$, which depends on the relative displacement, $\Delta y_j = y_{j+1} - y_j$, of these two masses and which is assumed to have the form

$$F_{interaction} = -K_j \Delta y_j + A_j(\Delta y_j)^2 + B_j(\Delta y_j)^3. \tag{8.1}$$

---

[*] Solitary waves are disturbances that retain their shape and speed as they propagate though a nonlinear medium via a fortuitous cancellation of the effects of dispersion and nonlinearity. Solitary waves that can collide and retain their shape are called solitons in recognition of this particle-like property. See chapter 3 on The Wave Equation and Other PDEs for examples and a more complete discussion.

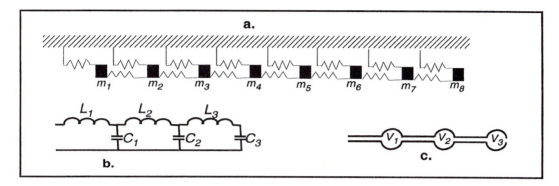

Figure 8.1: Systems that can be modeled by an interacting lattice: *a)* masses coupled by springs, *b)* a series of *LRC* circuits, *c)* volumes connected by tubes.

Two additional forces that do not depend on nearest-neighbor particle interactions are also present. The first attracts the mass toward the equilibrium position, $y_j = 0$,

$$F_{self} = -S_j y_j, \tag{8.2}$$

while the second is a velocity-dependent damping term:

$$F_{damp} = -C_j \dot{y}_j. \tag{8.3}$$

If we let each coupling constant, (K_j, A_j, B_j), act to the right, then Newton's second law applied to the j^{-th} mass can be written

$$
\begin{aligned}
M_j \frac{d^2 y_j}{dt^2} = & -K_j \Delta y_j + K_{j-1} \Delta y_{j-1} \\
& + A_j \Delta y_j^2 - A_{j-1} \Delta y_{j-1}^2 \\
& + B_j \Delta y_j^3 - B_{j-1} \Delta y_{j-1}^3 \\
& + C_j \dot{y}_j + S_j y_j \qquad j = \{1, 2 \cdots N\}.
\end{aligned} \tag{8.4}
$$

The A_j and B_j are the so-called cubic and quartic nonlinearities, respectively, due to their contributions to the potential energy of the system. Care must be taken to keep the cubic nonlinearity small so that the system does not become unstable due to a decrease in potential energy with increasing relative displacement.

Before beginning the discussion of properties that depend on the discrete nature of this system, it is instructive to show how a mass-spring lattice gives rise to a wave equation in the long wavelength limit.

8.2   Wave Equations and the Long-Wave Limit

If the phenomenon we wish to observe is large compared to the mass separation, a, then it is permissible to consider $y_j(t)$ to be a continuous function of x where

$x = ja$ and expand $y(x, t)$ in a second-order series by making the following substitutions:

$$\frac{M_j}{a} \longrightarrow \rho(x)$$

$$y_j(t) \longrightarrow y(x, t)$$

$$y_{j+1}(t) \longrightarrow y(x, t) + a\frac{\partial y(x, t)}{\partial x} + \frac{a^2}{2}\frac{\partial^2 y(x, t)}{\partial x^2}$$

$$y_{j-1}(t) \longrightarrow y(x, t) - a\frac{\partial y(x, t)}{\partial x} + \frac{a^2}{2}\frac{\partial^2 y(x, t)}{\partial x^2}$$

$$\dot{y}_{j+1} \longrightarrow \frac{\partial y(x, t)}{\partial t} + a\frac{\partial^2 y(x, t)}{\partial x \partial t} + \frac{a^2}{2}\frac{\partial^3 y(x, t)}{\partial x^2 \partial t}$$

$$\dot{y}_{j-1} \longrightarrow \frac{\partial y(x, t)}{\partial t} - a\frac{\partial^2 y(x, t)}{\partial x \partial t} + \frac{a^2}{2}\frac{\partial^3 y(x, t)}{\partial x^2 \partial t}. \tag{8.5}$$

We can now derive a number of standard wave equations by choosing restrictive values for the force constants K_j, A_j, B_j, C_j, and S_j. We show below that this leads to a wide variety of physically relevant systems in a mathematically compact and tractable model.

8.2.1   Classical Wave Equation

Assume a mass-spring lattice with identical masses, $M_j = M$, and Hooke's-law spring constants, $K_j = K$, set all other force constants equal to zero, $A_j = B_j = C_j = S_j = 0$, and substitute Eq. 8.5 into Eq. 8.4 to obtain the classical wave equation.

$$\frac{\partial^2 y(x, t)}{\partial t^2} = c^2\frac{\partial^2 y(x, t)}{\partial x^2}, \tag{8.6}$$

where

$$c = a\sqrt{K/M}. \tag{8.7}$$

8.2.2   Klein-Gordon Equation

Relaxing the previous conditions in order to allow non-zero self-spring constants, $S_j = S$, results in the Klein-Gordon equation,

$$\frac{\partial^2 y(x, t)}{\partial t^2} = c^2\frac{\partial^2 y(x, t)}{\partial x^2} + \mu y(x, t), \tag{8.8}$$

where $\mu = \sqrt{S/M}$. This equation, first derived in quantum field theory, is used to model the propagation of particles with non-zero rest mass and was discussed in chapter 3.[6]

It is, in fact, possible to approximate all the nonlinear Klein-Gordon equations treated in chapter 3 by a lattice of coupled oscillators, if we allow the self-spring constant, S_j to be a function of the relative displacement. A particularly intuitive system is a lattice of equally spaced pendula hanging from a horizontal rod and free to rotate with angle θ about the vertical. If the pendula are coupled by a torque proportional to their relative displacement $\Delta\theta$ and are also acted on by the vertical gravitational field which produces a torque proportional to $\sin\theta$, (i.e., let $x_j \rightarrow \theta_j$ and $S_j\theta_j \rightarrow \sin\theta_j$), then the *sine-Gordon* equation is obtained. The *kink/anti-kink* solution studied in chapter 3 corresponds to a rotation of the chain of pendula about their support axis. The pendula hang down at some point at the beginning of the chain, make a rotation over the top of the support rod, and then hang down for the remainder of the chain.

8.2.3   Korteweg-de Vries Equation

The cubic and quartic nonlinearites in Eq. 8.4 have played an important role in the development of nonlinear equations.[1,9] A higher-order expansion of the derivatives of $y_i(t)$ followed by a simple transformation of coordinates is all that is needed to produce the Korteweg-de Vries equation. This equation was first put forward in 1895 to model the propagation of water waves in a canal and has become the prototypical equation for the study of solitons.

8.3   Lattice Effects and the Short-Wave Limit

In order to see those features that are dependent on the discrete nature of the lattice, we shift our focus from partial differential equations to solving the N coupled ordinary differential equations of motion (Eq. 8.4), for each mass. Although the numerical simulation used in this program shows the solution for an arbitrary choice of coupling coefficients, we begin our study by making simplifying assumptions about the force constants, in order to obtain analytical solutions for the displacement.

8.3.1   Analytic Solution: Monatomic Lattice

The analytical solution for a chain of N uniform masses located at positions x_j and coupled with spring constants K is not difficult to derive. Rewriting Eq. 8.4, in a more convenient form, we obtain a set of N coupled differential equations with constant coefficients

$$M\frac{d^2y_j(t)}{dt^2} = K\big(y_{j-1}(t) - 2y_j(t) + y_{j+1}(t)\big) \quad j = \{1, 2 \cdots N\}. \tag{8.9}$$

These equations can be solved by assuming a sinusoidal solution

$$y_j(t) = A\cos(kx_j - \omega t + \delta_j), \tag{8.10}$$

where k is the wavenumber, ω is the angular frequency, and A is a constant amplitude. The mathematics is considerably simplified if we assume an exponential solution

$$y_j(t) = Ze^{i(kx_j - \omega t)} \tag{8.11}$$

and then take the real part in order to obtain the physical displacement. Substituting this equation into Eq. 8.9 and cancelling the complex amplitude Z and the $e^{i\omega t}$ term results in

$$-M\omega^2 e^{i(kx_j)} = Ke^{i(kx_{j+1})} - 2Ke^{i(kx_j)} + Ke^{i(kx_{j-1})}. \tag{8.12}$$

The x_j are uniformly spaced along the chain with separation a, so we can simplify by substituting $x_{j+1} = x_j + a$ and $x_{j-1} = x_j - a$ into the previous equation, setting $\omega_H^2 = 4K/M$, and cancelling the e^{ikx_j} term:

$$-\omega^2 = \omega_H^2(e^{ika} - 2 + e^{-ika})/4. \tag{8.13}$$

The exponential terms are now combined by using DeMoivre's theorem, $e^{i\theta} = \cos(\theta) + i\sin(\theta)$.

$$\omega^2 = \omega_H^2(1 - \cos(ka))/2 = \omega_H^2\sin^2(ka/2). \tag{8.14}$$

The desired solution is obtained by taking the real part of $y(t)$ and the positive root of ω^2.

$$y_j(t) = |Z|\cos(kx_j - \omega(k)t + \delta) \quad \text{where} \quad \delta = \arg z \tag{8.15}$$

$$\omega(k) = \omega_H|\sin(k/2)|. \tag{8.16}$$

Equation 8.15 is clearly a right-traveling sinusoidal wave for positive wave numbers and a left-traveling wave for negative wave numbers. Since k is arbitrary, there are in fact an infinite number of possible wavelengths. A general solution can be constructed using a Fourier integral (or sum if the medium is finite) of Eq. 8.15 over all possible wave numbers using the technique of Fourier analysis.

$$y_j(t) = \int_k A(k)\cos[kx_j - w(k)t + \delta(k)]\,dk. \tag{8.17}$$

What is most interesting is that ω is now a nonlinear function of k. Such functions were introduced in chapter 3 and are called dispersion functions. They are fundamental to the study of waves, since they predict not only the relationship between wavelength and frequency but also the group velocity, v_g.

At long wavelengths (i.e., small k) a Taylor expansion of the *sine* function shows that the phase and group velocity are equal to the velocity that was obtained for the classical wave equation, Eq. 8.7:

$$v_{phase} = \omega/k \quad = \omega_H a/2 \tag{8.18}$$

$$v_{group} = d\omega/dk = \omega_H a/2. \tag{8.19}$$

Wavelengths that are comparable to the lattice separation behave differently. First, the *sine* term on the right-hand side of Eq. 8.16 produces a cutoff at high frequency, ω_H. Unlike the classical wave equation, there is a maximum allowed frequency above which waves will not propagate. Second, the *sine* function is periodic, so it appears that more than one wavelength can produce the same frequency. This conclusion would be wrong. Sinusoidal waves that have the same frequency are in fact indistinguishable; they have the same wavelength as well as the same frequency. We will return to this topic after we have examined the effect of finite boundaries.

8.1   Exercise: Dispersion and the Klein-Gordon Equation
Repeat the above derivation but include a non-zero S_{self} spring constant in Eq. 8.9. Find the dispersion relationship and show that this dispersion relationship is equivalent to that of the Klein-Gordon equation.

8.3.2 Effect of Boundaries

As in the case of the classical wave equation, boundary conditions restrict the values of the wave number. Permissible values of k are called the eigenvalues of the system. For example, periodic boundary conditions require that the wave repeat itself every N atoms (i.e., the motion of the M_{j+N} mass is identical to the M_j mass.) This can only occur if the space occupied by a section of the lattice, $L = Na$, is an integer number of wavelengths. The wave number, k, in Eq. 8.15 thus takes on the following values

$$k_m = 2\pi m/Na \quad |m| = 1\ldots\infty, \tag{8.20}$$

where m is called the mode number. The sign of m will determine the direction of propagation of the wave on the chain.

Fixing two ends of a chain that are N atoms apart so that these ends cannot move imposes different restrictions. Both left- and right-traveling sinusoidal waves must be present and the wavelength must be a half-integer number of waves on the space $L = (N - 1)a$.

$$k_m = \pi m/(N - 1)a \quad m = 1\ldots\infty. \tag{8.21}$$

Notice that the change in wave number between modes is half that of Eq. 8.20. Only positive m values are allowed in our numbering scheme for Eq. 8.21, but the number of modes is the same as for Eq. 8.20 since periodic boundary conditions allow both positive and negative m.

Figure 8.2 illustrates many of the points discussed. It shows the dispersion function, Eq. 8.16, for a chain of 16 masses with periodic boundary conditions.

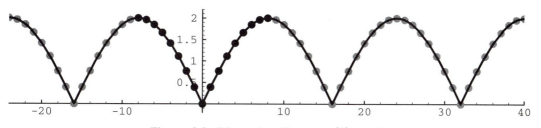

Figure 8.2: Dispersion Curve: $\omega(k)$ vs. k.

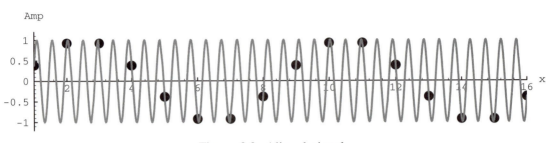

Figure 8.3: Aliased signal.

Angular frequency is along the ordinate and mode number (or wavenumber in units of $2\pi/L$), is along the abscissa. A dot is superimposed for each allowed mode. If this same graph were drawn for fixed boundary conditions, only positive mode numbers would be drawn but there would be twice as many allowed frequencies. Gray dots represent solutions that satisfy both the dispersion relationship and the boundary conditions but are indistinguishable from modes at longer wavelengths. This classification into groups of modes is discussed in section 8.3.3.

8.3.3   Brillouin Zones

The periodicity of a function can manifest itself in unusual ways. Consider a sinusoidal solution of a chain of 16 atoms with periodic boundary conditions. Substituting Eq. 8.20 into Eq. 8.15, we obtain an analytic expression for the initial position and velocity of the j^{-th} mass in the m^{-th} mode:

$$y_j(0) = \cos 2\pi m x_j/N \tag{8.22}$$

$$v_j(0) = \omega \sin 2\pi m x_j/N. \tag{8.23}$$

For large mode numbers the phase of the sinusoidal function in Eq. 8.22 can change by more than 2π from one atom to the next. The y_j values obtained using mode 34 of a periodic 16-atom chain are shown in Figure 8.3. The masses lie along a *sine* wave that has wavelength equal to half the length of the lattice. Clearly this is the same initial condition as mode 2. There are an infinite number of wavelengths in Eq. 8.22 that will fit the initial position of any mode. Consider a related problem in electrical engineering—recording a signal by sampling it at fixed time intervals. Information is lost and the signal is said to be aliased if it is sampled at too low a frequency. A general criterion (due to Nyquist) is that a signal must be sampled at over twice the highest frequency component to be accurately recorded. The lattice being used to sample the *sine* function, equation Eq. 8.22, has too low a spatial frequency to accurately reproduce the wave.

Short wavelengths are perfectly acceptable solutions to the dynamical equations. They satisfy the dispersion relation and the boundary conditions, but they are indistinguishable from the first N modes. Any contiguous group of N such modes forms a complete set and is referred to as a Brillouin zone. The modes $-N/2\ldots0\ldots N/2$ form the first Brillouin zone for the periodic lattice, while modes $1\ldots N$ form the first Brillouin zone for the fixed boundary lattice. Any Brillouin zone could be used to classify the modes, since they all mimic the modes in the first zone.

8.3.4 Diatomic Lattice

We conclude the discussion of analytical solutions by considering a chain with a unit cell of two atoms. Assume that all atoms on odd-numbered lattice points have mass M_1 while those on even numbered lattice points have mass M_2. These masses are coupled by linear force constants, K, and have separation, a, as before. The equations of motion for each type of mass can now be written

$$M_1 \frac{d^2 y_{2j-1}(t)}{dt^2} = K\big(y_{2j-2}(t) - 2y_{2j-1}(t) + y_{2j}(t)\big) \quad j = \{1, 2 \cdots N/2\} \tag{8.24}$$

$$M_2 \frac{d^2 y_{2j}(t)}{dt^2} = K\big(y_{2j+1}(t) - 2y_{2j}(t) + y_{2j-1}(t)\big) \quad j = \{1, 2 \cdots N/2\}. \tag{8.25}$$

We again assume exponential solutions

$$y_{2j-1} = A_1 e^{-ika} e^{i(2jka - \omega t)} \tag{8.26}$$

$$y_{2j} = A_2 e^{i(2jka - \omega t)}, \tag{8.27}$$

where the amplitudes, A_1 and A_2, will in general be different for different masses. Substituting Eq. 8.26 into Eq. 8.24, and Eq. 8.27 into Eq. 8.25 and cancelling the common exponential term results in

$$\omega^2 M_1 A_1 = \frac{K A_2}{a} \cos ka - \frac{2K A_1}{a} \tag{8.28}$$

$$\omega^2 M_2 A_2 = \frac{K A_1}{a} \cos ka - \frac{2K A_2}{a}. \tag{8.29}$$

These equations are consistent only if the following conditions are met:

$$\omega^2 = \frac{K}{aM_1} + \frac{K}{aM_2} \pm \frac{K}{a}\left[\left(\frac{1}{M_1} + \frac{1}{M_2}\right)^2 - \frac{4\sin^2 ka}{M_1 M_2}\right]^{1/2} \tag{8.30}$$

$$\frac{A_1}{A_2} = \pm \left[\frac{2K - a\omega^2 M_1}{2K - a\omega^2 M_2}\right]^{1/2}. \tag{8.31}$$

Notice that these equations predict two allowed frequency ranges or branches. Choosing the positive sign gives frequencies that are along the *optical* branch of the dispersion curve, while choosing the negative sign gives frequencies along the *acoustical* branch. The forbidden frequencies between these branches are known as a band gap.

8.4 *Numerical Methods*

In order to solve a system of ordinary differential equations for $y_i(t)$ using numerical techniques, we must discretize the time variable, $t_n = n\Delta t$. As in chapter 3, we use subscripts to denote the position of the atom in the lattice and superscripts to denote the time step of the calculation, $y_j^n = y_j(n\Delta t)$.

The second-order Taylor approximation (STA) is a numerically efficient algorithm for integrating the equation of motion of harmonic oscillator problems.[7] It is second-order in time, i.e., decreasing the time step by a factor of 2 increases the accuracy by a factor of 4, but, more importantly, the STA conserves the mechanical energy. The algorithm is very intuitive and can be written as follows:

$$y_j^{n+1} = y_j^n + v_j^n + \frac{1}{2}a_j^n\Delta t^2$$

$$v_j^{n+1} = \frac{y_j^{n+1} - y_j^n}{\Delta t} + \frac{1}{2}a_j^{n+1}\Delta t, \tag{8.32}$$

where the acceleration, a_j^n, is calculated using the right-hand side of Eq. 8.4.

8.5   Running CHAIN

8.5.1   A Quick Tour of CHAIN

Starting the program will initialize a Gaussian packet of 32 atoms with equal mass coupled by Hooke's law force constants (see Fig. 8.4). **Run** and **Pause** the

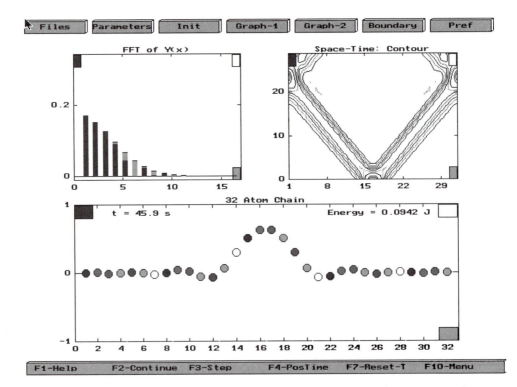

Figure 8.4: Opening screen with Fourier analysis and space-time contour plots.

simulation using the **F2** hot key. You may use **F7-Reset** to restore the initial conditions while the simulation is paused. Select other boundary conditions, such as **Boundary | Stationary | Fixed** or **Boundary | Stationary | Free**, and observe the behavior of the pulse.

You may zoom in or out by clicking the red scale button on the upper left-hand side of any graph and resetting xmin, xmax, ymin, and ymax. Zooming is possible only when the simulation is paused. Zooming changes the display without altering the data being plotted, the length of the chain, or the accuracy of the simulation.

Masses displayed on the screen can be dragged to new displacements by click-dragging while the simulation is stopped. If dragging is done at $t = 0$, the new displacement is stored as the initial condition so that it can be quickly reset. Double-clicking on an atom will bring up an inspector panel which allows the user to modify its position, mass, velocity, spring constants, and damping.

Change the number of atoms using **Parameters | Number of Atoms and Cells** and observe how this act affects the simulation. The dim vertical lines on the screen represet the cell boundaries and have no effect on the simulation. The user can insure that all unit cells have the same properties by selecting **Parameters | Examine Cell** and then copying the cell parameters to the entire chain. Many of the techniques discussed in chapter 3, section 3.5 on running the WAVE program, including tips for controlling execution speed, are applicable to CHAIN.

8.5.2   Inspector Panels/Mouse-Accessible Objects

The CHAIN program is built from encapsulated modules called *objects* which contain both data structures and the functions which act on the data. It is possible for the user to access several of these objects directly through the mouse, rather than through the menu. These mouse actions will generally bring up what we refer to as *inspector panels*, described below. If the CUPS utilities do not register a double click, you may need to adjust the double click time using the **Files | Configuration** menu option. If the computer registers a single click, as opposed to a double click, an inspector panel will not appear—rather, the single click allows you to drag an atom to a new position. Finally, if a single click cannot be associated with an atom, the program will display the coordinates where the click occured in the bottom left hand corner of the graph.

- Individual Atoms—Double-clicking the mouse on the desired atom brings up a panel for the dynamic variables, enabling the user to numerically modify an atom's y-position, mass, and velocity as well as the linear, quadratic, cubic, damping, and self coefficients in the equations describing the forces which act on the atom. Clicking **Make/Edit Source** displays an additional panel for external force variables. The user may choose a sine, Gaussian, symmetric pulse, pulse, modulated Gaussian, or user-defined function and set **Ampl**itude and **Freq**uency as desired. The following toggles are also present:

  - **Force/Displace**—specifies whether the *magnitude of the applied force* or the *y-displacement of the selected atom* will follow the source pattern. The **Displace** option creates, in effect, a boundary point: when a **Displace** source is selected, the atom's y-displacement will follow the selected pattern and the atom will reflect incident waves just like a a fixed (i.e., rigid) boundary.

- **Periodic/Single**—select a periodic pattern or a single pulse. You may set the time interval between periodic sources under the **Pref** panel on the menu.

If the computer registers a single click as opposed to a double click an inspector panel will not appear—rather, the single click allows you to manually adjust the atom's displacement by dragging the mouse to the desired position. Remember that if dragging is done at $t = 0$, the new displacement is stored as the initial position and can be restored by the **F7 Reset** hot key.

- Scaling—Clicking on the upper left (red) corner of any of the three graphs pulls up an inspector panel which enables the user to change the scale by choosing the minimum and maximum values for each axis. Automatic scaling with respect to either or both axes is also available (options **X**, **XY**, and **Y**). Because data are stored in arrays independent of scale, changing the scale does not affect the data being plotted or result in a "loss" of data which lies off the visible portion of the graph.

- **Expand/Contract Graph**—The upper right (purple) corner of each graph acts as a toggle between a full-screen and smaller (default mode) graph.

- Display preferences can be set by clicking the lower right (blue) corner of the main (atom) graph. You may find that changing these preferences noticeably alters the simulation speed. For example, it takes less time for the computer to draw outlined atoms than to draw atoms and fill them with color. In general, the simpler graphs are the fastest to draw. The following options are available:

  - Select **colored atoms** or **outlined atoms**.
  - Define atom radii in terms of pixels or in terms of the horizontal scale. (One unit on the horizontal scale represents the distance between the centers of adjacent atoms.)
  - **Scale Atom Size By Mass** or display atoms with uniform radii.
  - **Show velocity** of each atom as a vector emanating from its center.
  - **Show envelope** of each atom's oscillation.

- Attributes of graphs 1 and 2 (the analysis graphs) can be set by clicking on the lower right (blue) corner of the appropriate graph. All types of plots may be "frozen" so that they remain unchanged during continued simulation. Additional options are available for the following plots:

  - **Displacement: Y(t), Velocity:V(t), Phase Space: V(y), Y2(t) vs. Y1(t)**—change the display to any one of these four graphs.
  - **Space Time-Contour**—display legend of colored isoclines, choose the palette, or change display to **Space Time-3D**.
  - **Space Time-3D**—specify the Euler angles describing the orientation of the plot (allows you to rotate the plot), hide/show the box defined by the three axes of the plot, or change display to **Space Time-Contour**.

- **FFT-y(x)**—select Fourier transform of y-displacement, $y(x)$, or velocity, $v(x)$; choose a basis set composed of sine terms or a basis set containing both sine and cosine terms (a basis set of sine terms is generally used to analyze simulations which fix the endpoints at $y = 0$.); hide/show maximum values of the Fourier power spectrum.
- **FFT-y(t)**—choose the number of points from which to construct the Fourier transform; choose a basis set composed of sine terms or a basis set containing both sine and cosine terms; hide/show maximum values of the Fourier power spectrum.

8.5.3 Menu Items

- **Files**

  - **About CUPS**—Information about the Consortium for Upper-Level Physics Software (CUPS).
  - **About the Program**—Information about CHAIN.
  - **Configuration**—Set a directory for temporary storage of overflow files, change the screen color, or check available memory.
  - **Load**—Load an ASCII file containing a configuration of oscillators from a drive.
  - **Save**—Allows you to save the current configuration of oscillators to a drive. Files are saved in standard ASCII format and can be edited with any text editor. Note: the parameter and display preferences will not be saved.
  - **Save as**— Specify a filename and save the current configuration as above. Up to two lines of comments may be included in the file. These comments will be displayed when the file is loaded.
  - **Exit**

- **Parameters**

  - **Number of Atoms and Cells**—Set the total number of atoms and the number of atoms per cell. A cell is simply an arbitrary grouping of atoms that can be modified as a whole and used to initialize periodic conditions.
  - **Examine Atom**—Edit data pertaining to the chosen atom. This option is equivalent to double-clicking on an atom (see section 8.5.2).
  - **Examine Cell**—Edit data pertaining to all of the atoms in a given cell, and apply the resulting configuration to a single cell or to the entire chain. The **Apply to Entire Chain** option makes it easy to set up periodic conditions.
  - **Time**—Set the time and change the time increment between calculations. This option is useful when something interesting happens and you wish to re-examine a particular time interval. Change the time to 0 and run the simulation as desired. Resetting the graph will return the configuration displayed when you set $t = 0$. You can use the **Time** option to speed up or slow down the simulation by adjusting the time steps between calculations, but keep in mind that larger time steps compromise numerical accuracy. Speed can be adjusted without affecting accuracy by selecting **Pref→Speed** or by using the hot keys **F5 Slower** and **F6 Faster** while the simulation is running

- **Scale Chain**, **Scale Graph1**, **Scale Graph2**—Adjust the scale of the graph of the chain, the upper left graph, or the upper right graph, respectively. These options are equivalent to double-clicking on the upper left (red) corners of the appropriate graphs.

- **Init**—Specify initial conditions.

  - Set up the oscillators in one of the following configurations: **zero**, **random** displacement and velocity, **sine** wave, **Gaussian** distribution, **Modulated Gaussian** (a sine wave multiplied by a Gaussian), **Pulse**, **Symmetric Pulse**, **Modes**, or **User-defined function**.
  - **Center of Mass Frame** sets the initial linear momentum of the system to zero. This keeps the image from "floating" off the screen in periodic or free configurations for which total momentum would not otherwise be zero.
  - **Zero Velocity** sets $v_i = 0$ for all atoms without disturbing displacement. This feature allows you see what happens when kinetic energy is removed from a system of oscillators.

- **Graph-1**, **Graph-2**—Disable the graph, clear the graph, or choose from the following plots:

  - **Displacement: Y(t)**—Display the y-displacements of up to five atoms as a function of time.
  - **Velocity: V(t)**—Display the velocities of up to five atoms as a function of time.
  - **Source Power: P(t)**—Displays, as a function of time, the instantaneous power delivered by a source.
  - **Energy: E(t)**—Display the energy of the system as a function of time.
  - **Phase Space: V(y)**—Depict the motion of up to five atoms in a phase space defined by velocity and vertical displacement.
  - **Y2(t) vs. Y1(t)**—Plots the y-positions of the two user-specified atoms, Y1 and Y2, on the x- and y-axes, respectively.
  - **Space Time-Contour**—This graph displays isoclines representing points of equal amplitude as a function of position (on the x-axis) and time (on the y-axis).
  - **Space Time-3D**—Rather than using isoclines to represent amplitude on a two-dimensional graph, this option generates a three-dimensional depiction of amplitude as a function of position and time.
  - **FFT-y(x)**—This Fourier power spectrum represents the Fourier transform of y-displacement as a function of x-position. This graph may slow down the simulation noticeably, because it is updated with each screen redraw. The red bars show the most recent transform, while the blue bars store maximum amplitudes obtained by each Fourier component.
  - **FFT-y(t)**—This Fourier power spectrum represents the amplitude of a selected atom as a function of time. This graph is blank during an initial period of data collection.

- **Boundary**

  - **Stationary**—the default mode is a stationary reference frame with **Fixed** boundaries. **Free** and **Periodic** boundaries are also available for stationary reference frames. The **Periodic** option is equivalent to placing a free oscillator duplicating the right-most atom's behavior at the left of the chain, and a free oscillator duplicating the left-most atom's behavior at the right of the chain. **Periodic** boundaries are often used to approximate an infinite one-dimensional lattice, since the wave propagates freely from the right hand side to the left hand side of the chain. Note that creating a "displacement" source has the effect of setting a fixed boundary (see section 8.5.2).

  - **Moving**—a moving reference frame allows the user to follow a window of the chain as the oscillations propagate. The user sets the velocity of the moving reference frame.

- **Preferences**

  - **Display**—This option is equivalent to clicking the lower right corner of the chain graph (see section 8.5.2).
  - **Color**—Color by atom, color by cell, or alternate the color. Be careful not to change the color of the atoms to the background color of the graphs.
  - **Files**—Set drive and directory path for overflow files.
  - **Speed**—Allows you to speed up the animation by increasing the number of time-step calculations carried out before the screen is redrawn, or to slow down the animation by inserting a Pascal **PAUSE(t)** command after every screen redraw. Because the time step, Δt, between calculations remains unchanged, the accuracy of the underlying numerical algorithm is not affected by this option.
  - **Sources**—set the period of periodic sources. Sources are set as **periodic** or **single** when their respective atoms are edited.
  - **Demo**—when **Demo Mode** is enabled, a demonstration simulation will evolve for the amount of time specified by the parameter **DemoTime**. Then the simulation will reset and the demo will be run again. **Demo Mode** models whatever simulation—including graphs, sources, etc.—is present when enabled.

8.5.4 Hot Keys

At the bottom of the screen you will find a list of the hot keys and their functions. Note that the functions of the hot keys change when the simulation is running. The following hot keys are available when the simulation is not running:

- **F1 Help**—No help is available if the help file is not in the selected directory path. To change directory path, go to **Pref | Files**.

- **F2 Run**—Run simulation.

- **F3 Step**—Proceed to the next screen redraw The number of time steps, Δt, calculated between screen redraws depends on the speed of the simulation.

- **F4 PosTime/NegTime**— This key acts as a toggle between positive and negative time steps, causing the simulation to run forward or backward in time.

- **F7 Reset**—Restore the initial configuration of $t = 0$.

- **F10 Menu**—Escape to the menu at the top of the screen.

  The following hot keys are available when the simulation is running:

- **F1 Help**

- **F2 Pause**

- **F3 Show-V, Hide-V**—This key toggles between showing and hiding the velocity vectors of the atoms.

- **F4 PosTime/NegTime**—This key acts as a toggle between positive and negative time steps, causing the simulation to run forward or backward in time.

- **F5 Faster**—Increases the speed of the simulation *without affecting numerical accuracy* by increasing the number of time steps between each redrawing of the screen.

- **F6 Slower**—Decreases the speed of the simulation by adding (or lengthening) a Pascal **PAUSE[t]**command between time steps.

8.6   Exercises

8.6.1   Simple Systems

The simplest possible chain is a harmonic oscillator, HO, consisting of a single mass connected to two fixed boundaries through springs. Although the HO has only a single frequency, it is one of the most important problems in physics. Adding a second atom introduces the possibility of interaction. We will briefly study these simple systems in order to learn how to add and control driving forces and other parameters in preparation for the study of longer driven chains. *Note:* Be prepared to slow down the simulation when modeling systems with small numbers of atoms.

8.2   **Inspecting Atom and Cell Parameters**
You can change parameters for a single atom by **Parameters | Examine Atom** or by double-clicking on a mass to bring up the atom's inspector panel and editing the appropriate field. You can examine a group of atoms by selecting **Parameters | Examine Cell**, although this method does not allow you to add forcing functions to atoms. The **Examine Cell** method does, however, allow you to quickly change the entire chain by copying

the parameters from one cell into all other cells. Both methods allow the user to examine and edit coupling constants such as the Hooke's law force,

$$F_j = -K_j \Delta y_j = -K(y_j - y_{j+1}) \tag{8.33}$$

and add quadratic and cubic forces. These terms will be used to study nonlinear effects in a later subsection. The linear, quadratic and cubic forces act between the selected mass and the mass on the right. The self spring force, $-S_i y_j$, is proportional to the displacement from the equilibrium position, and friction is proportional to the negative of the velocity, $-C v_j$.

a. Type CHAIN at the DOS prompt to begin the program and see if the program loaded properly by clicking **F2-Run** key. Reset the initial conditions by clicking **F7-Reset**.

b. Click on an atom to the right of the initial Gaussian pulse and change its mass to 10. Run the program and notice that the wave "bounces" off the heavy mass.

c. Examine the four-atom cell containing the heavy mass from the previous exercise. Set all the masses in this cell to 4. Notice that the chain shows the heavy masses as larger circles. If too many large masses begin to obscure the display, turn off mass scaling by clicking on the lower right (blue) corner of the chain graph and changing the display options.

d. Use the blue attributes button of the chain graph to turn off the **Scale Atom Size By Mass** option and turn on the **Show Velocity** option. Run the program to see how these changes affect the display.

8.3 Adding a Source Term

A source term can be added to any atom's equation of motion by checking the source button in the atom's inspector panel. The inspector subpanel should appear with additional options that let you specify source type (**sine, Gaussian, symmetric pulse, pulse, modulated Gaussian** or **user defined**) and source parameters (**amplitude, frequency, force/displace,** and **periodic/single**). Note the source frequency is in units of cycles per second, rather than angular frequency, ω, units of radians per second. A check box asks if the source function specifies a driving force or displacement. Select **amp** $= 1$, $\omega = 1.5$ and the **Sin, periodic,** and **Force** radio buttons. To deactivate the source, reselect the inspector by double clicking on the mass and select the **Disable** button.

a. Set **Init ⏐ Zero** and add a default Gaussian source term to an atom near the middle of a chain of 32 atoms. Run the program. Does the source atom act like a fixed boundary or can the wave pass through the source atom?

b. Change the source term from **Displace** to **Force** and repeat a.

8.4 Simple Harmonic Oscillator

Set the boundary conditions to **Fixed** and set the number of atoms equal to 3 and the atoms per cell equal to 3 using **Parameters ⏐ Number of Atoms and Cells**.

a. Click and drag the center mass toward the top of the screen and run the program. Since there are two springs with spring constants attempting to restore equilibrium, the angular frequency, ω, should be equal to one if the default values $M = 1$ and $K = 1/2$ have not been changed. Measure the value of ω.

b. Double-click on the center mass and create a sinusoidal source with frequency $f = 0.14$. A graph should be visible showing the source as a function of time. Select the **Force**—Don't select the **Displacement**—mode and then click **Exit**. Select **Graph-1 | Displacement: Y(t)** and **Graph-2 | Source Power: P(t)** and run the program. Observe the beating of the particular and the complementary solutions of the HO differential equation. Notice in the source power graph that the source does not always add energy to the system. Sometimes energy is removed depending on the phase angle between the force, $F(t)$, and velocity, $v(t)$. What should the beat frequency be?

c. Use the atom inspector to add a frictional force, $F_{friction} = -Cv$ to the mass by setting $C = 0.2$. Run the program again and observe that the system quickly settles into a steady state.

8.5   Oscillator Resonance

Predict the resonant frequency and resulting steady state amplitude with $M = 1$, $C = 0.5$, and $K = 1$. Check your prediction against the simulation.

8.6   Power Delivered to a Simple Harmonic Oscillator

The average power delivered by a source as a function of the frequency is given by

$$P(\omega) = \frac{F^2}{M} \frac{C/2M\omega^2}{(\omega^2 - K/M)^2 + 4(C/2M)^2\omega^2}, \tag{8.34}$$

where F is the amplitude of the sinusoidal driving force, C is the damping, and K is the spring constant. Simulate the simple harmonic oscillator, in which $M = 1$, $C = 0.2$, $K = 4.0$, and $F = 1.0$. Use the **Energy vs. Time** and **Source Power vs. Time** graphs to examine the source power delivered. Check your results against the analytical expression. The blue, red, and green curves in the **Energy vs. Time** graphs show kinetic energy, potential energy, and total energy, respectively.

8.7   Coupled Oscillators

Create four masses with fixed boundaries and set the spring constant coupling mass two to mass three equal to 0.1.

a. Drag mass 2 away from its equilibrium position and run the program. The amplitude of mass two, $y_2(t)$, will decrease with time while that of mass three will increase. After approximately three oscillations almost all the oscillation will have been transferred to mass three, after which time the process will reverse itself and the energy will be transferred back to mass two. Set **Graph-1** to generate a **Displacement:Y(t)** graph that shows the motion for both atoms. Freeze the graph. Now change the coupling constant to 0.5 and generate another **Displacement:Y(t)** in **Graph-2**. Compare the graphs and check them against theoretical predictions.

b. Set **Graph-1** to generate an **Displacement:Y(t)** graph and **Graph-2** to generate an **Y1(t) vs Y2(t)** graph. What type of figure do you observe in **Graph-2**?

c. Use **Parameters I Examine Cell** to examine all four atoms. Set the initial positions of the two middle atoms to 0.5 and the initial velocities to 0. Run the simulation with the same analysis graphs as in the previous exercise. How is the motion different? Measure the frequency of oscillation and compare to theory.

d. Repeat b but set the initial positions of one of the middle atoms to 0.5 and set the other to -0.5. Run the simulation. How is the motion different from b? Measure the frequency of oscillation and compare to theory.

The motion observed in b or c is referred to as a normal mode of oscillation and the frequencies of these modes are called characteristic frequencies or eigenfrequencies.

8.8 Coupled Oscillator Resonances

If the undamped coupled oscillator system introduced in the last exercise is driven with a force $F(t) = F_0 \cos \omega t$ then both y_2 and y_3 will have steady state solutions with the following time dependence, where y_2 is the amplitude of the forced atom:

$$y_2(t) = \frac{F_0}{M} \frac{\omega_0^2 + \omega_c^2 - \omega^2}{(\omega_0^2 - \omega^2)(\omega_0^2 + 2\omega_c^2 - \omega^2)} \cos \omega t \qquad (8.35)$$

$$y_3(t) = \frac{F_0}{M} \frac{\omega_c^2}{(\omega_0^2 - \omega^2)(\omega_0^2 + 2\omega_c^2 - \omega^2)} \cos \omega t, \qquad (8.36)$$

where

$$\omega_o = \sqrt{(K/M)} \quad \omega_c = \sqrt{(K_c/M)}. \qquad (8.37)$$

a. Given a coupling constant from mass two to mass three of $K_c = 0.1$, and using default values for all other parameters, use a spreadsheet or similar tool to plot the amplitudes of the two center atoms, y_2 and y_3, as a function of driving frequency from $\omega = 0.5$ to 1.0. Notice that two resonances occur at the eigenfrequencies determined in the previous exercise.

b. Add a small amount of damping to remove transients and drive the circuit close to the first resonance by making atom two a source. Compare the amplitude to theory. Notice that the two masses move in phase with each other.

c. Drive the circuit close to the second resonance and notice that the two masses move out of phase with each other. You will see this phase relationship between adjacent atoms again when you study band gaps on a diatomic lattice.

d. There is a point midway between the two resonances where the amplitude of y_2 becomes zero. Drive the system at this frequency to test if the motion is indeed zero.

Wave Number and Frequency

This subsection introduces measurement tools and visualization techniques for properties such as wavelength, frequency, eigenfunctions (i.e., modes), and dispersion.

8.9   Single-Mode Behavior

The motion of a single mass can be very complicated if the system is placed into an arbitrary initial condition as shown in Figure 8.5a. There are, however, certain initial conditions where each mass executes simple harmonic motion about its equilibrium position, as shown in Figure 8.5b. Such an initial condition is called a mode. A mode is a very special type of collective motion since, when the chain is placed in such a state, each mass oscillates in simple harmonic motion with the same frequency, ω. Modes are often countable. You may have studied modes on a string or in an organ pipe in your introductory physics course. Modes on a chain are similar but their number is finite if we restrict our counting to the first Brillouin zone as explained in 8.3.3.

a. Start the program and change both boundaries to **Fixed** using **Boundary | Stationary**. Now select the **Init | Modes** option. Choose mode 4 in the input screen and click **OK**. All 32 masses will lie along a two-cycle sine curve. Run the simulation and let the system evolve. Observe the standing wave in the chain graph. The upper left-hand corner contains a fast Fourier transform of the positions of the masses, **FFT of Y(x)**. The upper right-hand corner is initially blank but will show a **Space-Time Contour** after 32 updates of the chain graph. Remember that an update is performed only after the number of calculations specified under **Pref | Speed** by the number of time steps between screen redraws. The analysis plots can therefore have coarse time resolution even though the actual calculations done on the chain are fairly accurate.

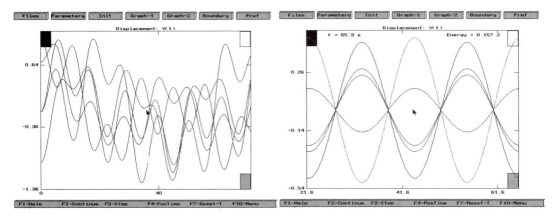

Figure 8.5: Displacement of five atoms on a 32-atom chain as a function of time: **a** Multiple modes, **b** Single mode.

b. Select **Graph-1 | Displacement: Y(t)** and click on five different masses. Graph-1 will display $y(t)$ for these masses when the program runs. Select **Graph-2 | Phase Space: V(y)**. This type of plot is often used in mechanics. The angle between the origin and the point (x, v) gives the phase angle of a mass as it executes simple harmonic motion about the equilibrium point. Run the program a number of times after choosing different mode numbers from **Init** and describe the behavior of the displacement and phase space plots. The interested reader may wish to examine the the CUPS program ANHARM by Bruce Hawkins for a more detailed discussion of its use in oscillator dynamics.

c. Set the initial condition to be a Gaussian and repeat b. Notice that the motion appears quite a bit more complex than when the system was in a single mode.

8.10   Fourier Analysis

The **FFT** analysis in the previous exercise did not produce a graph with a single non-zero bin, even though the user selected a single mode as an initial condition. It is possible for the bin number to match the mode number if the user selects the correct number of atoms. Remember that the FFT algorithm requires that the number of data points be a power of two. Also remember that the Fourier sum repeats itself after N grid points. The last atom on the computer screen (i.e., atom 32—the fixed point on the right-hand side) is not the last lattice point in a *sine* cycle but the first atom in the next cycle!

a. In order for the number of atoms in a mode to be a power of 2 with fixed boundaries you should change the number of atoms to 33 using **Parameters | Number of Atoms and Cells**. Reset the analysis graph to **FFT-y(x)** after you set the number of atoms and then select mode 4. The **FFT** should now show a single non-zero bin. Why does the amplitude of the mode oscillate with twice the period of the standing wave?

b. For periodic boundary conditions the number of atoms should be equal to a power of 2 in order for each **FFT** bin to correspond to a mode. Change the number of atoms to 32, set the boundary to be periodic, and select an FFT analysis plot to check this.

c. Another way to create a system with fixed boundaries is to fix a mass so that it cannot move on a system with periodic boundaries. Make sure the number of atoms is 32 and select periodic boundaries and an **FFT-y(x)** analysis graph. Double click on mass one and make it a displacement type source with zero amplitude. Select **Init | Sine** and set the wavelength to 16. Run the program and notice that only bin 2 is non-zero in the **FFT** even though the initial condition was set for mode 4 for fixed boundaries. Set the wavelength to 64 and notice that the **FFT** graph again does not display a single non-zero bin. What complete set of eigenfunctions is being used with periodic boundary conditions? With fixed boundary conditions?

The **FFT** analysis graph actually displays a *power spectrum* as described in chapter 3. It uses different sinusoidal basis sets depending on the boundary conditions and combines terms with the same spatial frequency into a single coefficient. If fixed boundaries are selected it uses sine functions that have zero crossings at the boundaries. Functions that have integer and half-integer wavelengths on L are allowed. If periodic boundaries are selected the algorithm uses sine and cosine functions and combines coefficients that have the same frequency.

Notice that the red bar in the **FFT** graph shows the current value of the **FFT** while the blue bar records the maximum value.

d. Fourier analysis can be used in the time domain as well as the spatial domain. Reinitialize the system with fixed boundaries and place the system in mode 4. Select **Graph-1 | FFT Y(t)** and run the program. A time series FFT will appear after 256 screen updates. Since a sine-cosine basis set is used, the displayed power spectrum has half this number of bins. Each bin represents a different temporal frequency in the displacement function of the selected mass, $y_i(t)$. What are the frequencies of the first bin and the last bin? What is the difference in frequency between bins?

8.11   Free-Fixed System

The boundary condition for a free end is that the derivative of the function be zero:

$$y_N(t) = 0. \tag{8.38}$$

a. Write an expression for the allowed wavelengths, λ, for a chain that is fixed at one end and free at the other end. Assume the chain has N masses and that the free end is an anti-node.

b. Set the left boundary to fixed and the right boundary to free and place the system in a fixed-free eigenstate using **Init | Sine**. Observe the motion.

8.12   Dispersion

You have seen that the motion of each mass is harmonic if the system is in a single mode. You will now examine an important difference between these modes. With fixed boundary conditions, set the mode number to 1 and measure the wavelength, λ, by using a mouse-down in the chain graph. Measure the period, T, of the mode using a mouse-down to measure a time interval from a displacement, $y(t)$, plot in **Graph-1**. Another way to measure the period is by recording two simulation times from the chain view when the masses are at successive maximum displacements. To get accurate readings, you may need to toggle the time step between forward and reverse and adjust the animation speed. Convert λ and T to wave number, k, and angular frequency, ω:

$$k = 2\pi/\lambda$$
$$\omega = 2\pi/T. \tag{8.39}$$

Record these values and calculate the phase velocity, c.

$$c = \lambda/T = \omega/k \tag{8.40}$$

Repeat the above measurement for mode 8. Notice that the phase velocity, c, has changed. This simple exercise points out a fundamental difference between waves on a one-dimensional chain of coupled masses and waves satisfying the classical wave equation. The phase velocity depends on the wavelength.

8.13 Boundaries and Dispersion

Change the boundary condition by selecting **Boundary | Stationary | Periodic**. Now select mode 1 and measure the wavelength, λ, and the period, T. Repeat for mode 4. Notice that mode 4 for periodic boundaries has the same phase velocity as mode 8 for fixed boundary conditions.

8.14 Standing Waves

The Chain program can generate standing waves using periodic boundary conditions if the user specifies two modes, m_1 and m_2 such that $m_1 = -m_2$. Make sure that **Periodic** boundary conditions are selected and then click **Init | Modes**. Enable mode 2 and the user definable mode and type -2 into the data field next to the user-defined modes and then click **OK**. Select **Phase Space: V(y)** plots for **Graph1** and **Graph2**, choosing two masses separated by half a wavelength. Run the program and notice that the phase space vectors, $(x(t), v(t))$, for any two masses are either in phase or exactly π radians out of phase. This phase relationship will be true only if the traveling wave components, m_1 and m_2, have equal amplitude.

8.15 Galloping Waves

Bring up the **Init | Mode** panel and create left- and right-traveling waves with amplitudes 0.5 and 1.0. Go to **Pref | Display** or click the attributes button in the lower right-hand corner of the chain plot and set the envelope to true. Run the simulation and observe that the waves do not produce perfect interference. Turn the envelope off and run the program so that you can observe the motion of the peaks. Notice how the peaks remain stationary for most of the cycle and then move very rapidly to a new position? These phase peaks can actually go faster than the speed of light![3]

8.16 Equivalence of Brillouin Zones

Find positive and negative modes that are equivalent to the $M = 2$ mode for a chain with periodic boundary conditions. Run the program and demonstrate these modes.

Dispersion on a Monatomic Lattice

The dispersion relationship for a lattice of point masses, M, acted on by Hooke's law nearest-neighbor forces, K, was derived previously as

$$\omega(k)^2 = \omega_H^2 \sin^2(ka/2) \quad \text{where} \quad \omega_H = 2\sqrt{K/M}. \tag{8.41}$$

This subsection will examine some consequences of this relationship, particularly at short wavelengths.

8.17 Dispersion Function

Use a spreadsheet or similar program if one is available to do the following repetitive calculations.

a. Plot the dispersion function (Eq. 8.41) using the default values $K = 1/2, M = 1$.
b. What are the allowed wavelengths and frequencies for a lattice of 32 masses with periodic boundary conditions? Create a data table of T, ω, λ, and k. (See blank sample table at the end of this section.)
c. Set both lattice boundary conditions to fixed and measure the frequency $\omega(k)$ for the six modes $(1, 2, 4, 8, 16, 32)$. Compare to the analytic results to b.

8.18   Beats

Choose **Boundary l Periodic** conditions and then enable modes 4 and 6, using **Init l Modes**. Run the program and notice that the masses on the chain exhibit a beat pattern. Notice also that since the phase and group velocities are different, the beat node and the waves within the envelope travel at different speeds. Select a space-time contour analysis graph and explain how phase and group velocity manifest themselves on this type of graph.

8.19   Group Velocity

Derive expressions for the phase and group velocities from Eq. 8.41. Plot these expressions using a spreadsheet or similar package if one is available. For what wavelengths do waves on this lattice behave like classical waves? For what wavelength is the group velocity half the phase velocity? What is the closest mode number for which $v_g = v_p/2$? Place the system in two modes on either side of this mode number and observe the beats. Compare the measured phase and group velocities to the analytical expressions.

8.20   Critical Points

If you use the second Brillouin zone you can compare two modes that lie on either side of a maximum in Eq. 8.41. Place the system in such a state and describe the motion. What is the phase velocity? The group velocity?

8.21   Fixed Boundary

Examine beats on a fixed boundary lattice. Use a chain of 65 atoms and modes 20 and 24. Notice that nodes of the beat envelope pass right through the fixed atoms. It is even possible for the envelope function to have a maximum next to the boundaries.

A Driven Monatomic Lattice

Assume you have a long chain of coupled oscillators that is being driven at one end with a sinusoidal displacement:

$$y_1(t) = D \cos \omega t. \tag{8.42}$$

We will examine the subsequent motion of the chain subject to different boundary conditions on the right hand side.

8.22   Driven-Fixed Chain

The steady state for a chain driven at one end and fixed at the other will be a standing wave that satisfies Eq. 8.42 and $y_N = 0$.

a. Show that the steady-state solution is a standing wave with a wavelength-dependent amplitude, $A(k)$:

$$A(k) = \frac{D}{\sin kL}. \tag{8.43}$$

The amplitude will be infinite whenever the driving frequency hits a resonance, $\sin kL = 0$. Demonstrate this effect. Hint: Remember that the length of an N-atom chain is $N - 1$.

b. In order to remove transients, add a small amount of damping, $B_i = 0.01$, to each atom on the chain using **Parameters | Examine Cell**. Set the right-hand boundary to fixed and the left-hand boundary to free. Make the mass on the left boundary a sinusoidal source. Make sure this source is set to displacement and periodic. Examine the amplitude for frequencies close to and far from resonance. In order to determine when steady state is reached, set one of the analysis graphs to record mass displacements, $y(t)$, and the other to record energy, $E(t)$.

8.23   Impedance Matching

Although it is not possible to eliminate all reflections at the end of a chain, it is possible to absorb waves at a single frequency by a judicious choice of mass, M_N, and damping coefficient, C_N, for the last atom. Such a choice of boundary conditions is known as impedance matching.

a. Show analytically that the following conditions on the last atom will eliminate reflection of a sinusoidal wave from that boundary:

$$M_N = \frac{M}{2} \tag{8.44}$$

$$C_N = \frac{K}{\omega} \sin ka. \tag{8.45}$$

b. Construct a driven-free monatomic chain of 32 atoms using the default values of M and K. Set the sinusoidal source to have a frequency of $f = 0.1$ Hz and impedance match the free end by setting the last atom to have $M_N = 1/2$ and $C_N = 0.633$. Run the simulation. Do you see evidence of a reflected wave? Recalculate the damping coefficient and repeat for a source frequency of 0.2 Hz.

c. Change the source to be a modulated Gaussian and repeat b. How effective is impedance matching of this type of pulse?

d. Set the width parameter on a Gaussian source so that the maximum amount of this pulse is absorbed at the other end. Run the program and estimate how much was absorbed by observing the energy.

8.24   Evanescent Waves

The dispersion function (Eq. 8.41) predicts that there are frequencies for which there are no corresponding wave numbers. It is, of course, possible to grab hold of the chain and shake it at any frequency whatsoever. What type of motion is induced if the driving frequency is above the high frequency cutoff, ω_H? The answer is surprisingly simple; the dispersion relationship is still valid but you must use complex values for the wavenumber, $k = \pi/a - i\kappa$.

 a. Calculate ω_H for a chain with default values of mass and spring constants.
 b. Consider a chain that is infinite to the right of the source. Show that a complex wave number leads to the following relationship for the displacement of an atom to the right of the source

$$y_j = De^{-\kappa ja} \cos(\omega t - j\pi), \tag{8.46}$$

where $\kappa = 2/a \cosh^{-1} \omega/\omega_H$.
 c. Set up a driven-free system of 32 atoms and add a small amount of damping to every atom on the chain in order to remove transient effects ($C_i = 0.01$ works well). Drive the system at a frequency that is 1 percent higher than the cutoff frequency. Set the mass of the last atom to $M/2$ but do not add any more damping. What do you observe? After the transients have died out you may want to turn on the **Show Envelope** feature using the blue attributes button and observe the shape of the amplitude.
 d. Select a source power analysis plot. Is power being delivered to the chain?

8.25   **Damped Chain**
Add a damping coefficient of $C_i = 0.2$ to every atom of an impedance matched chain and drive the system at 0.1 and again at 0.2 Hz. Explain similarities and differences between the observed motion and that of the evanescent wave.

8.26   **Tunneling**
Build a 64-atom chain that has eight atoms with mass $M = 4$ starting at position number 25. Add damping of 0.01 to each atom to remove transient response. Drive the system at the left boundary with a frequency of 0.113 Hertz and impedance match the right boundary. Run the simulation until steady state is reached, and then turn on the envelope option. Describe the types of motion in the three regions of the chain.

Dispersion on an Anchored Lattice

In exercise 1 you derived the dispersion relationship for a lattice of point masses acted on by Hooke's law nearest neighbor forces, K, and non-zero self-spring forces, S.

$$\omega(k)^2 = \frac{S}{M} + 2\frac{K}{M}(1 - \cos(ka)) = \omega_L^2 + \omega_H^2 \sin^2(ka/2) \tag{8.47}$$

This system clearly exhibits both a high frequency and a low frequency cutoff. The low frequency cutoff is, in fact, very similar to the dispersion relationship for the propagation of electromagnetic waves in a plasma.

$$\omega(k)^2 = c^2 k^2 + \omega_p^2 \tag{8.48}$$

The plasma frequency accounts for the reflection of radio waves by the earth's ionosphere below a certain frequency.

8.27 **Anchored Lattice Dispersion**
Use a spreadsheet or similar package to do the following calculations.

a. Plot the dispersion function (Eq. 8.47) using the values $K = 1/2, M = 1, S = 1$.
b. Create a data table of T, ω, λ, and k for all the even modes of a 32 atom chain. (See blank sample table at the end of this section.)
c. Select **Parameters | Examine Cell** and set the self-spring constant to 1 for all atoms in the first cell. Apply the first cell's values to the entire chain. Set both lattice boundary conditions to fixed and measure the function $\omega(k)$ for the five modes $(1, 4, 8, 16, 32)$. Compare to the analytic results of b.

8.28 **Evanescent Waves**
Compare the behavior of evanescent waves below the low frequency cutoff to those above the high frequency cutoff. For best results, add some damping to remove transient response and impedance match the unforced boundary. Notice the phase angle between neighboring atoms.

Dispersion on a Diatomic Lattice

In order for a diatomic lattice to exhibit modal behavior, both the dispersion relationship (Eq. 8.30), and the amplitude relationship (Eq. 8.31), must be satisfied.[5] Unfortunately, the **Init | Modes** menu assumes a monatomic chain. We can, however, construct the correct initial conditions using parser input under **Init | Unser Defined Function**.

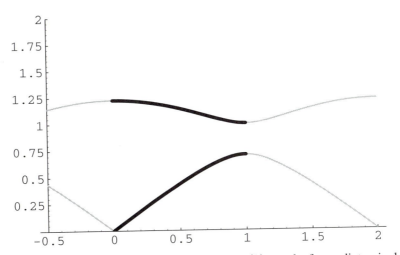

Figure 8.6: Acoustical and optical dispersion curves, $\omega(k)$ vs. k, for a diatomic lattice. Wave number is in units of $\pi/2a$.

8.29   **Using The Parser**

Use the default configuration for a 32 atom monatomic chain to test the parser.

  a. Enter the following formula into the parser text field and run the program.

$$0.5 * \sin (Pi * x/8). \tag{8.49}$$

Is the chain in a single mode? What are the modes?

  b. The parser will set the initial velocity of the atoms if time is an explicit parameter in the input formula. To test this enter the following function for the initial condition.

$$0.5 * \sin (Pi * x/8 - 0.276 * t). \tag{8.50}$$

  c. Calculate the wave number and angular frequency for the fifth mode and test your result by entering into the parser the correct formula for this mode.

  d. Set the number of atoms per cell to 2 using **Parameters | Number of Atoms and Cells** and then use **Parameters | Examine Cell** to set the two masses in the first unit cell to 1 and 2. Apply this condition to the entire chain. Select mode 2 using **Init | Modes**. Is this initial condition for a monatomic chain a good approximation of a mode for the diatomic chain?

  e. Repeat d but select mode 5. Examine one of the lighter atoms using a displacement graph. What evidence do you see that the system is not in a single mode? How many modes do you think there are?

8.30   **Diatomic Lattice Dispersion Function**

Using a spreadsheet or similar tool, plot the dispersion curve (Eq. 8.30 for a diatomic lattice with $M_1 = 1$ and $M_2 = 2$. You should obtain a graph similar to Figure 8.6. Make sure to include both the acoustical and optical branchs of the function. What is the high frequency cutoff? What two frequencies define the band gap?

8.31   **Acoustical Modes**

The easiest way to write a parser function that satisfies Eqs. 8.26 and 8.27 is to multiply a sinusoidal function by a second function that alternates from 1 to $1 + \delta$ between neighboring atoms. One such amplitude modulation function is $1 + \delta| \cos \pi x/2|$. Enter the following initial condition for a diatomic chain after setting the masses to be $M_1 = 1$ and $M_2 = 2$:

$$0.5 * \sin (Pi * x/8 - 0.225 * t)$$

$$*(1 + 0.028 * Abs(\cos (Pi * x/2))). \tag{8.51}$$

Notice that the amplitude ratio is 1.028.

  a. Observe the motion. In what mode or modes is the system? Notice that this function is very close to the initial condition for a monatomic lattice which explains why **Init | Modes** produces a single acoustical mode for long wavelengths.

  b. What are the wave number and the angular frequency? Compare the results of the simulation with Eqs. 8.30 and 8.31.

8.32 **Optical Modes**

The atomic motion in optical branch excitation is quite different from an acoustical branch excitation. Enter the following initial condition for a diatomic chain after setting the masses to be $M_1 = 1$ and $M_2 = 2$:

$$0.5 * \sin (Pi * x/8 - 1.20 * t)$$

$$*(1 - 1.487 * Abs(\cos (Pi * x/2))). \qquad (8.52)$$

Notice that the amplitude ratio is -0.487.

a. Run the simulation and observe the motion. How would you number this mode?

b. What are the wave number and the angular frequency? Compare the results of simulation with Eq. 8.30.

c. What qualitative difference do you observe between acoustical and optical modes? Explain why an ionic crystal, i.e., a crystal with alternating positive and negative charge, is easily excited by an electromagnetic wave if the field oscillation matches a frequency on the optical branch but not if the field oscillation matches a frequency on the acoustical branch.

8.33 **Higher-Order Modes**

The shorter wavelength modes don't look quite as sinusoidal as the long wavelength modes. Modal behavior can still be seen by examining displacement and FFT-y(t) analysis plots of the atoms.

a. Use a spreadsheet or similar package to plot the amplitude ratio, Eq. 8.31, as a function of frequency.

b. Program a calculator or spreadsheet to evaluate Eqs. 8.30 and 8.31 for an arbitrary mode number. Modify the wave number, frequency, and amplitude ratio in the traveling wave formula from the previous exercise to place the system into what would be mode 5 on the monatomic lattice. Test your new initial condition by performing a temporal FFT on the displacement of an atom.

c. There are two modes at the zone edge, since a Brillouin zone edge occurs in a diatomic system when $k = \pi/2a$. Study the motion of the system in each of these modes. You may have to modify the functional form of the parser string to input one of these modes.

8.34 **Multimode Distributions**

Set the following conditions: 32 atoms, two atoms per cell, and **periodic** boundary conditions. Edit a cell so that the masses of its atoms are one and two; apply this cell condition to the entire chain. Set one analysis graph to be the displacement of two atoms of different masses and set the other to be a FFT of Y(t) analysis graph and select one of the smaller masses for analysis. Set the initial conditions using **Init | Mode** for the following exercise.

a. Run simulations for mode 1. Does this initial condition place the system primarily in the acoustical or optical branch? Run the system for mode 12. Did this initial condition place the system primarily in the acoustical or optical branch?

b. How can Eq. 8.31 be used to explain the differences between modes 1 and 12?

c. Use pencil and paper to predict the relative amplitudes of the optical and acoustical excitation if mode 12 is selected.

8.35   Driven-Fixed Diatomic

Only a minor modification needs to be made to the formulas to impedance match a diatomic lattice with alternating masses M_1 and M_2 that is driven at one end and fixed at the other. If the chain has an even number of masses, then the last atom should have the following mass and damping coefficient:

$$M_N = \frac{M_2}{2} \tag{8.53}$$

$$C_N = \frac{A_1}{A_2} \frac{K}{\omega} \sin ka, \tag{8.54}$$

where the amplitude ratio at an allowed frequency can be found using Eq. 8.31.

a. Create a chain having free boundaries and 32 masses alternating between 1 and 2. Set the left boundary to be a sinusoidal source with $f = 0.1$ Hz. Set the right boundary to be half the normal mass and give it a damping coeficient of $C_N = 0.438$. Run the program. Enable the the **Show Envelope** feature after steady state has been reached and describe the motion. Which atoms have the greater displacement?

b. Repeat at a frequency of 0.18 Hz and a damping coefficient of 0.785. What is different about this motion?

8.36   Band Gaps

Create a system with alternating masses and free boundaries as before and calculate the frequences that bound the bandgap. Create a sinusoidal displacement source at the left boundary with a frequency of $f = 0.13$ and add a small amount of damping to every atom in the system to remove transients. Run the simulation and describe the behavior. Is the wave evanescent?

Nonlinear Phenomena

Fermi, Pasta, and Ulam (FPU) performed the first numerical study of a system of coupled nonlinear oscillators at Los Alamos in 1955 using the MANIAC I computer.[2] They studied a chain of 32 coupled oscillators—just as you are doing—and their goal was to study the relaxation of energy from a single mode into thermal equlibrium with all other modes in the system in accordance with the equipartition theorem. This equipartition of energy did not occur and although it is now possible with 20/20 hindsight to provide an analytical basis for this null result, the FPU problem is still one of the classic heuristic uses of the computer.

We will employ their original technique—albeit with better graphics—to analyze the problem.

8.37   Mode Mixing

Create a 33 atom chain with fixed boundaries and then use **Parameters | Examine Cell** to produce a system with the following parameters:

- Mass: $M_i = 1$
- Linear coupling: $K_i = 1$
- Quadratic coupling: $A_i = 0.25$

Set an analysis graph to produce a FFT of displacement and initialize the system in mode 1 with amplitude 1.

a. Run the system and observe that the nonlinearity begins to distort the initial sinusoidal waveform.
b. Expand the FFT analysis graph to full screen to speed up the program. You may also want to increase the time-step using **Parameters | Time** if your computer is slow. Continue to run the program and watch the energy redistribute itself into the higher order modes. However, after about 158 periods of the fundamental mode, the energy has reconcentrated itself into mode 1! This revival of the initial mode occurs again and again as the system evolves as shown in Figure 8.7
c. Is the effect observed in b a fluke? Will the system exhibit revivals if it is started in other modes?
d. Will other parameter values for the initial mode amplitude, the quadratic coupling coefficient, or number of atoms produce different results?
e. Do revivals occur with cubic coupling?

8.38   Kink Solitons

Create a 64 atom chain with default parameters and increase the time-step using **Parameters | Time** to $\Delta t = 0.2$ for added speed. Set the linear coupling coefficient to one 1 but keep all nonlinear coupling coefficients zero for now. Set the left and right boundary to be free. Now use double-clicks on the first and last atoms to make them into sources. Set the source type to be **User Defined** and enter the following function into the parser string field for the first atom:

$$-1.25 * (1 + \tanh(0.52 * t - 4)). \tag{8.55}$$

Enter zero, 0, into the parser string field for the last atom. Change the scale on the chain graph so that the minimum value of y is -3.0 using the red scale button.

a. Run the program until the edge of the step reaches the center of the chain and then pause the program. Use **Boundary | Moving** to enable a moving reference frame with $v_{ref} = 1.00$. Continue to run the program and notice that the edge spreads. You may need to adjust the velocity from time to time but eventually the initial wave will fill

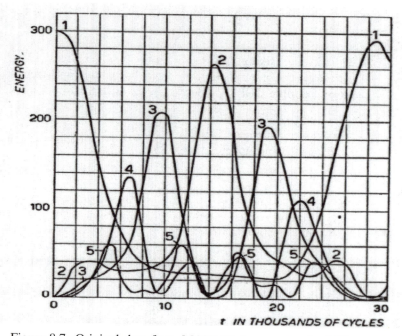

Figure 8.7: Original data from MANIAC I computer for FPU problem.

the entire 64 atom chain. Clearly the step did not retain its shape due to the different phase velocities of the various Fourier components.

b. Reset the program to $t = 0$ using **F7** and set the quadratic coupling coefficient to $B = 0.2$ for every atom in the chain. Turn off the moving refence frame and repeat a. Turn on the moving reference frame when the step is near the center of the screen. Notice that the step moves with a group velocity $v = 1.04$ and does not change its shape. The nonlinearity has canceled the dispersion for this particular pulse shape. This type of pulse is an example of a *soliton* and is called a kink.

c. Set the source on the right hand side to be $-1.25 * (1 + \tanh(0.52 * t - 4))$ and run the program. Notice how the two solitons collide without changing their shape.

d. The amplitude, S, and speed, Ω, of a kink soliton are related in the following manner:[9]

$$y(x,t) = -|S|\frac{\sqrt{K}}{4A}\left[1 + \tanh\frac{\Omega t - Sx}{2\sqrt{K}}\right] \qquad (8.56)$$

$$\Omega^2 = KS^2 + S^4/12 \qquad (8.57)$$

Verify this relationship through simulation.

8.7 Advanced Projects

8.39   Triatomic Lattice

If there are P atoms in a primitive cell for a one-dimensional chain, the dispersion curve will split into P branches. Set the unit cell to have three atoms and the masses of these atoms to be $M_1 = 1$, $M_2 = 2$, and $M_3 = 3$. Demonstrate that there are three branches. How many of these branches would you characterize as acoustical?

8.40 Long-Range Forces

The interactions in CHAIN.PAS are all nearest-neighbor forces. Investigate the effect of relaxing this condition. Reprogram CHAIN to include interaction among next-nearest neighbors. How do these interactions effect the dispersion curve?

8.41 Ergodic Behavior

Put yourself in Fermi, Pasta, and Ulam's shoes. Can you add a nonlinearity to the chain that will lead to an equipartition of energy among all of the modes?

		THEORY						SIMULATION				
	wave-length	wave number	period	angular freq.	w/k	dw/dk	wave-length	wave number	period	angular freq.	w/k	dw/dk
2												
4												
6												
8												
10												
12												
14												
16												
18												
20												
22												
24												
26												
28												
30												
32												
64												

(row label: M O D E S)

Figure 8.8: A sample table for data collection.

Acknowledgments

We would like to thank Laurence S. Cain for making numerous useful suggestions about the manuscript and for testing the software.

References

1. Dodd, R.K. Eilbeck, J.C, Gibbon. J.D., Morris, H.C. *Solitons and Nonlinear Wave Equations.* London: Academic Press, 1982.

2. Fermi, E. *Collected Works of E. Fermi, vol. II.* Chicago: University of Chicago Press, 1965.

3. Harter, W.G. Galloping waves and their relativistic properties. *American Journal of Physics* **53**:671, 1985.

4. Ingard, K.U. *Fundamentals of Waves and Oscillations.* New York: Cambridge Univ. Press., 1988.

5. Kittel, C. *Introduction to Solid State Physics*, 5th ed. New York: John Wiley and Sons, 1976.

6. Sakurai, J.J. *Advanced Quantum Mechanics.* Reading, MA: Addison-Wesley, 1967.

7. Stanley, R.W. Numerical methods in mechanics. *American Journal of Physics* **52**:499, 1984.

8. Visscher, P.B. *Fields and Electrodynamics.* New York: John Wiley and Sons, 1988.

9. Zabusky, N.J. Computational synergetics and mathematical innovation. Journal of Comparative Physics, **43**:195, 1981.

Appendix

Walk-Throughs for All Programs

These "walk-throughs" are intended to give you a quick overview of each program. Please see the Introduction for one-paragraph descriptions for all programs.

A.1   Walk-Through for DIFFRACT Program

The default program is HUYGHENS: Slits. The diffraction pattern due to a single slit is plotted on a viewing screen placed close to the slit.

- **Click the mouse at other locations up the viewport.** This will plot the pattern on a viewing screen placed at that position.

- **Change the XY Scale Multiplier using the slider.** It is then possible to see the pattern at a range of distances from the slit. You can examine the manner in which the pattern changes over a wide range of distance.

- **Change the number of slits** using the buttons.

- **Change the wavelength using the slider.**

- **Vary the slit width and slit separation (if you have more than one slit).**

Now choose HUYGHENS: Point Sources. The default screen shows the intensity pattern in the plane of two radiating sources as a contour plot.

- **Change the distance between the sources.**

- **Change the phase relationship between the sources.** Note the way in which the lobes change.

- **Use the Threshold slider to allow the pattern to be viewed at larger distances from the sources.**

- **Select a different number of point sources.** Note that one point source gives the expected spherical distribution. Also note how the lobe structure changes as the number increases.

The second Menu option is **CORNU**. As well as studying some of the features of the spiral itself you can also study diffraction at an edge, at a single or double slit and at a single or double obstacle. Choose the **Double Slit** option. The viewport shows the diffraction pattern for the default parameters indicated by the set of sliders.

- **Change the left-hand slit width.** The pattern shown is for two slits of identical width.

- **Change the slit separation.**

- **Change the distance of the source from the slit and the distance of the screen from the slit.** Make the slit width small and the slit separation large. Observe the manner in which the pattern changes. You can compare this result to that obtained with the HUYGHENS: Slits program.

- **Vary the wavelength.**

- **Vary the relative widths of the two slits.**

The third Menu option—**FRAUNHOFER**—looks at Fraunhofer diffraction from a slit or set of slits, from a rectangular aperture and a circular aperture. Choose the Single and Multiple Slits option. The default screen shows the diffraction pattern under Fraunhofer conditions for two slits. It is possible in this simulation to superimpose different patterns for comparison purposes. The **F3-Clear All** hot key allows you to clean the viewport when required. The **F2-LastPlot** hot key replots the most recent simulation.

- **Vary the number of slits using the buttons.**

- **Change the slit width and slit separations.**

- **Vary the wavelength.**

Choose the **Rectangular Aperture** option. The default screen shows a three-dimensional plot of the intensity and the contour plot.

- **Vary the dimensions of the rectanglar aperture.**

- **Vary the wavelength of the incident light.**

- **Use the mouse in the dummy screen to reposition the viewing screen.** Drag the small red square to a new postion and rerun the program. The threshold slider allows you to enhance to low intensity features.

The final Menu option—**COHERENCE**—looks at some issues in partial coherence and fringe visibility. Three configurations can be studied - two sources and a single circular aperture or slit, two sources and two slits, and a line source of finite width and a two slits. Two further options allow you to study the operation of the Michelson stellar interferometer for measuring the separation distance in a double star and the diameter of a star. Choose the option **Two Slits-One Line Source**. The lower viewport in the default screen shows the intensity variation on a viewing screen. The upper viewport the visibility function.

- **Make the slit width a minimum.** This removes the effect of finite slit width.

- **Vary the distance between the slits.** Note how the visibility changes. Find a slit separation that reduces the visibility to zero. The upper viewport shows the variation of the visibility function with $d\Delta\theta/\lambda$. The green line is set at the correct value and the values of visibility and $d\Delta\theta/\lambda$ are displayed. Direct measurement of the visibility is possible, using the mouse in the lower viewport.

- **Vary the wavelength of the source.**

Now choose the **Star Diameter** option. The default screen shows the intensity variation in the stellar interferometer.

- **Increase the separation of the mirrors from a low value.** Find a setting for zero fringe visibility. Can you find another setting?

You might like to look at a few problems. Try HUYGHENS: Point Sources, Problem d. This simulates a phased radar array. As a second and third problem, look at COHERENCE: Two Slits-One Line Source, Problem b and also COHERENCE: Star Diameter, Problem a, where you can see if you successfully can measure the diameter of the stars Arcturus and Aldebaran.

A.2 Walk-Through for SPECTRUM Program

The default program is GRATINGS: Transmission Grating-Spectrum. The default screen shows the spectrum for a grating whose parameters are set by the sliders. Two other simulations are available under this Menu item—Transmission Grating-Resolution and Blazed Reflection Grating. Choose the third option. The opening screen contains two viewports. The upper viewport shows the orders of diffraction and the blazing envelope. The lower viewport shows the effect of the blazing envelope.

- **Reduce the blaze angle.** Note the way in which the blazing envelope can be adjusted to select a particular order. Set the blaze angle so that the first order is selected using the zoom control for fine adjustment. Note that the blaze wavelength is the same as the wavelength of the source.

- **Vary the angle of incidence.** See how the selectivity is maintained.

- **Adjust the steplength.**

- **Adjust the wavelength.**

The second menu option is **PRISM**. You can examine the spectra produced by a series of light sources and prism materials and also study the ability of the instrument to resolve wavelengths. Choose the **Prism-Spectrum** option. An input screen allows you to choose the prism material and the light source from default sets or you can choose your own.

- **Choose hydrogen as the source and crown glass as the prism material.** The screen shows the spectrum in the viewport. Use the **Zoom Range** button to display the spectrum—the Balmer spectrum. Note the ray paths in the lower viewport.

- **Adjust the angle of incidence.** Set the angle of incidence to show total internal reflection of some rays.

- **Press the F5-Data hot key to display the angles through the prism.** The angles are defined in the text.

- **Press the F9-Defaults hot key to show the input screen.** Choose a dense flint prism and the mercury source. See that it is possible to easily resolve the mercury doublet at small angles of incidence. This can also be studied using the **Prism-Resolution** program.

The third Menu option—**MICHELSON**—looks at features of the Michelson interferometer. You can study the spectrum, measure refractive indices of solid, and gases and also look at the basis of Fourier transform spectroscopy. Choose the **Spectrum** option. An input screen allows you to select two wavelengths, the effective plate separation, and the distance that the moving mirror is to travel in an animation option. The opening screen shows two viewports. The upper viewport shows the intensity as a function of position across a viewing screen, in false color. The lower viewport shows the combined intensity.

- **Press the F3-ShowRings hot key to display the rings structure.**

- **Press the F2-ClearRings hot key to redisplay the intensity pattern.**

- **Select the Animation button.** The single viewport shows the manner in which the intensity changes as the moveable mirror travels.

- **Reselect your initial conditions with two wavelengths that are closer together.**

Note that the input screen offered the option of a problem. This is also the case with two other options under this category. Two problems are accessible via hot keys in Fourier transform spectroscopy.

The final option—**FABRY**— looks at the operation of the Fabry-Perot interferometer. You can study the spectrum and also look at the question of resolution. An input screen allows you to choose a pair of wavelengths, the etalon separation and plate reflectivity. The opening screen shows a pair of viewports similar to the case of the Michelson interferometer.

- **Use the mouse to expand the angular range.** Click on a new low and a new high angle. The range will change in both viewports. Note the shape of the lines.

- **Change the reflectivity.** Rezoom the angular range and note the effect.

- **Change the etalon separation.** Note the way in which the rings system changes, using the **F3-ShowRings** hot key.

- **Change the wavelength separation.**

Problems you might like to look at could include the Blazed Reflection Grating: Problem b, the Michelson interferometer-Spectrum: Problem c, and the Fabry-Perot interferometer: Problem b.

A.3 Walk-Through for WAVE Program

The initial system is a standing electromagnetic wave. Energy density and space-time analysis graphs are enabled.

- **Press the F2 hot key** to start the simulation. Press F2 to pause the simulation after the space-time contour plot appears. Press F7 to reset the initial conditions. Repeat but wait for the second rendering of the space-time contour plot. Notice the compression along the time axis.

- **Expand and then contract a graph** using the purple expand button located in the upper right hand corner of any of the graphs.

- **Mouse-down inside a graph** and notice the coordinate readings in the left hand corner.

- **Examine the attributes** of each graph by clicking the blue attribute button in the bottom right hand corner of each graph. Set the energy density graph attribute to show magnetic field energy and run the simulation.

- **Create a segment** with a different index of refraction by clicking on the green segment button labeled "n" in the bottom graph and then clicking inside the graph. Double click on the green segment to bring up the segment inspector and change its width to 0.25. Close the inspector and drag the segment to the right hand side of the medium. Use **Init | Gaussian** to create an initial disturbance on the left hand side of the medium. Run the program and observe the multiple reflections in the system.

Each of the following sample data files demonstrates a different aspect of the program. All data files end in a *.WAV extension. Load these files using **Files | Open** and run the simulations. Note: Poorly configured DOS machines without enough memory may not be able to initialize all the analysis graphs.

- **FABRY.WAV** simulates an 11 segment Fabry Perot interference filter with transmission maxima at 3 Hz. Detector graphs are enabled to allow the user to examine the transmitted and reflected waves. Notice the precursor, i.e., transient, to the steady state. Interference filters don't work in optics for short laser pulses!

- **BEATS_KG.WAV** presents a very nice example of dispersion. Note how different phase and group velocities manifest themselves on the space-time contour plot.

- **QM_SHO.WAV** simulates a Gaussian wave packet propagating in a simple harmonic oscillator potential.

- **LASER.WAV** demonstrates that a laser cavity will usually oscillate in multiple longitudinal cavity modes. We use this demonstration with an optical spectrum analyzer and a He-Ne laser in junior lab. Increase the number of FFT data points using the blue attributes button for best effect.

- **BION.WAV** is a good introduction to particle field theory. This simulation shows the production of a bound state after the collision of two soliton-like pulses.

Read the book! The first four exercises introduce various aspects of the visualization and analysis tools used in WAVE.

A.4   Walk-Through for CHAIN Program

The initial system consists of a 32 atom chain with a Gaussian displacement profile. FFT and space-time analysis graphs are enabled.

- **Press the F2 hot key** to start the simulation. Press F2 again to pause the simulation after the space-time contour plot appears. Press F7 to reset the initial conditions and repeat but wait for the second rendering of the space-time contour plot. Notice the compression along the time axis.

- **Expand and then contract a graph** using the purple expand button located in the upper right-hand corner of any of the graphs.

- **Examine the attributes of each graph** by clicking the blue buttons in the bottom right-hand corner of each graph.

- **Mouse down inside a graph** and notice the coordinate readings in the left-hand corner.

- **Double click on a mass** to bring up its attributes inspector. Change the mass to 4 and close the inspector. Use **Init | Zero** to zero the positions. Drag the center mass up using the mouse. Run the simulation.

Each of the following sample data files demonstrates a different aspect of the program. All data files end in a *.CHN extension. Load these files using **Files | Open** and run the simulations.

- **COUP_OSC.CHN** simulates the coupled oscillator problem through the use of a four mass chain with fixed ends.

- **OPTICAL.CHN** simulates a traveling wave in the optical branch of the phonon dispersion curve.

- **ACOUSTIC.CHN** simulates a traveling wave in the acoustical branch of the phonon dispersion curve.

- **TUNNEL.CHN** simulates tunneling of a wave through a segment of the chain where the wave is evanescent.

- **BAND_GAP.CHN** simulates the response of a diatomic chain when it is driven at a frequency within the bandgap.

- **SOLITON.CHN** shows how nonlinear coupling between atoms can be used to balance dispersion, thereby producing a soliton.

Read the book! The first three exercises show the user how to use the program to set up and analyze the simple harmonic oscillator using CHAIN.

A.5 Walk-Through for FOURIER Program

. The initial screen comes up with the default waveform, which is a square wave.

- **Click repeatedly on the F3-Step hot key (by mouse or keyboard).** You will see the result of successive waves being added to produce a synthesized wave form.

- **Select the F2-Run/Stop hot key, after adding four or five harmonics.** Observe how the synthesized wave form approaches the square wave more and more closely, as increasing numbers of harmonics are added.

- **Select the menu item Fourier giving the list of waveforms that may be used, and select the choice Input Function.**

- **Enter any expression involving known functions of the variable x on the input screen.** For example, you might try entering simply x, or if you wanted, you could try such a poorly behaved function as $\sin(x)/x$, which the program should be able to handle.

- After the function is accepted, either step through successive harmonic additions (using the F3 hot key), or else add them automatically, using the F3-Run/Stop hot key.

- Select either 1-D DFT or 2-D DFT from main menu (one and two-dimensional discrete Fourier transforms), and you will be given a number of menu selections to choose from.

- For example, select 2-D DFT, and then select Import Picture.

- Select NEWPICT.TXT as the data file to input, using the mouse or arrow keys. Be sure that the full name of the file appears in the selection window, even if you need to eliminate the file path in order to do so. After the program reads in the data file, it displays the pixel intensities both as a contour map, and also, less meaningfully, as a three-dimensional display.

- Select the Power/Gray F8 hot key, and then select "gray levels of real part," and the program will display the intensities of each pixel using a gray scale, thereby producing a crude photograph.

- Select the F4-Transform hot key, and the program will compute the two-dimensional discrete Fourier transform. Note how the result is dominated by low frequency components (at the center of the three-dimensional display).

- Select the F4-Inv Trans hot key, and an input screen will appear that allows you to apply a filter before doing the inverse transform. For example, select the low pass filter by mouse click, and leave the pixel radius set at its default value (15), which means that the filter excludes pixels more than 15 away from the central pixel. After the program computes the inverse transform, the display is again presented in the form of three-dimensional and contour plots.

- Select the F6-Power/Gray hot key again to produce the photograph-type display. Note, how the image is more fuzzy than the original, owing to the exclusion of high frequency components.

A.6   Walk-Through for RAYTRACE Program

Click the mouse or press return to clear the credits screen to start the **Fermat** option.

- **Fermat.** This option is to illustrate Fermat's principle of least path for refraction and to show its agreement with Snell's law of refraction. A ray of light is traced from one corner of a square region until it meets a boundary which separates material of refractive index n from material of refractive index n'. At the boundary the path of a ray to the opposite corner is shown, and the path of a ray obeying Snell's law is also shown. The optical path for the route from one corner to the opposite corner is calculated. The optical path for this route depends upon the position at which the ray meets the boundary. Graphs are drawn which show the optical paths for rays to all positions along the boundary. The total optical path from corner to corner is also shown plotted against the position at which the ray might cross the boundary.

It is possible, by the use of **Sliders**, to move the position at which the ray meets the boundary, and a vertical line on each graph indicates this position. Below the diagram of the ray path the numerical values change as the sliders are moved, and it will be found that the separate calculations for Fermat's principle and Snell's law show that Snell's law is in agreement with Fermat's principle as the values of $n \sin i$ and $n' \sin i'$ are the same at the position at which the minimum optical path occurs, and that at this point there is just one ray above the boundary.

There are sliders that make it possible to change n and n', the refractive indices of the two regions, and also the vertical position of the boundary separating the two regions. When any of these three sliders is adjusted the graphs of optical path are redrawn automatically.

- **Select the menu item Mirage.** In this simulation of a mirage use is made of the reversibility of rays of light to show which rays of light could arrive at an observer's eye after "reflection" at the hot surface of a road. The simulation takes the eye of the observer to be at a height of 1.5 m above the road. Light from the sky will be seen by the observer to illuminate the road. Immediately in front of the observer this will be diffuse and the surface will appear dry. At some distance from the observer the rays of light from the sky will be deviated into a curve so that, at some point, they suffer total internal reflection and the mirage of a "wet" surface is formed. A **Slider** is provided so that the direction of the ray at the observer's eye can be varied. A hot key is available to change the function setting the variation of refractive index with height above the road surface.

- **Next select the menu item Fiber.** Use the **Sliders** provided to investigate the factors that have lead to the adoption of optical fibers around 1 to 10 μm in diameter. The sensitivity to small changes in a refractive index for a single mode 8 μm step index fiber is quickly appreciated by trying to make suitable adjustments to the sliders for the fiber index and for the cladding index. The 1 to 90 ns pulse width shows the overlap that can occur due to the variation in path lengths for multi-mode fibers. The scale on the display of the output pulses is automatically adjusted to show the region of interest.

- **The menu item Rainbow** leads to a pull-down menu. Select either **Primary rainbow** or **Primary and secondary rainbow**. The laws of reflection and refraction are applied for rays incident on a raindrop and the deviation of these rays is plotted against the impact parameter which can be controlled by the **Slider**. The change in direction of the ray is plotted from the observer's viewpoint, so that the variation of the angle of the rays to the direction of the incident rays is given, which equates to 180° deviation. Below the graph of impact parameter versus ray angle, there is a graph showing frequency of deviation against ray angle. This second graph effectively shows the intensity of light against ray angle, and shows how different colors have maximum intensity at different angles to the incident illumination. The arc of the rainbow is produced by this constant angle for each color being rotated around the axis of the direction of the incident rays.

- **The menu item Lenses** also leads to a pull-down menu. **Select Object at Infinity.** The rays from a point object are traced through the refracting surfaces of the lens design to show the aberrations. If the data file TESSAR.LEN is in the chosen directory, it will be loaded when the **Lenses** option is selected for the first time. If the file TESSAR.LEN is not available, data for a thin meniscus lens will be used. At the top the rays are shown emerging from the lens and they will usually be seen to converge to an image point and then diverge. The display gives the value of the focal length and the back focal distance. There is a **Slider** to move the position of a viewing screen along the axis. The spot diagram at the lower right shows the intersection of the rays with the viewing screen. The display at the bottom left gives a detailed view of the lens being investigated. The **hot key F5-Zoom In/Zoom Out** controls the magnification of the spot diagram. The **hot key F3-Center up** re-centers the spot diagram so that the ray passing through the center of the screen is near the center of the display. The **hot key F3-Center up** also redraws the slider with a smaller range so that finer adjustments of the position of the image screen can be made.

- **Object Near Lens.** In this option the position of a point object can be set on an input screen that appears on use of the **hot key F8-Object.** The rays for a range of object positions can be investigated.

- **Edit Lens Design.** On selecting this option the user is presented with an input screen which contains the details of the design of the lens currently being simulated. The number of refracting surfaces and aperture stops can be specified along with the positions of the surfaces their radii of curvature and the refractive index of the spaces in between.

Exercises

- **Reducing spherical aberration.** The amount of spherical aberration when aperture and focal length are fixed varies with lens shape. If the spherical refracting surface is thought of as consisting of a large (infinite) number of prisms arranged around the axis of the lens it can be seen that the deviation of a ray depends on the orientation of the prism to the incident ray. The incident ray will undergo minimum deviation when it makes the same angle to the surface normal as the emerging ray does to the surface normal. Examine the spherical aberration for a plano-convex lens with the sperical side away from the focal point and towards the focal point. Aim for a focal length of about 5 cm.

A.8   Walk-Through for QUICKRAY Program

Access to the ray diagrams is provided under the main menu title **QuickRay**.

- **Use the mouse or keyboard to select QuickRay followed by Ray Diagram for a Lens.** You see a ray diagram in which the object is represented by a yellow vector and the image by a white vector. The blue lines represent the three principal rays that run from the object towards the observer. Numerical values are given for the object and image distances and the focal length of the lens. These can be used by an instructor who wishes to show that they always satisfy the equation

$$\frac{1}{\text{image dist}} + \frac{1}{\text{object dist}} = \frac{1}{\text{focal length}}. \qquad (A.1)$$

  Use the mouse to capture and drag the object. See what happens when the object is brought inside the focal length of the lens. See what happens when the object is moved to the other side of the principal axis or the other side of the lens. Note that the object distance is always positive (real object), whereas the image distance is positive for a real image and negative for a virtual image. When you are done, return the object (approximately) to its original position.

- **Click on the F4-Focus hot key.** Use the mouse to capture and drag the focal point of the lens. (The "active" focal point is the one marked with a cross and is initially to the left of the lens.) See what happens when the active focal point is moved to the other side of the lens. The displayed focal length becomes negative to indicate a diverging lens. **Click on the F2-Object hot key** to enable the mouse to capture and drag the object again. Does the "real is positive" convention apply for a diverging lens? (Answer: Yes.)

- **Click on the F3-Lens hot key.** Use the mouse to capture and drag the lens and its focal points as a unit. When you are done, **Click on the F10-Menu hot key** to recover the main menu.

- **Select QuickRay followed by Ray Diagram for a Mirror** and repeat the above exercises. Note that, unlike the lens case, when the object is moved to the other side of the mirror the sign of its focal length changes (because a two–sided reflecting surface that is concave to the left (positive focal length) is *convex* to the right (negative focal length). Does the "real is positive" convention apply here? (Answer: Yes.)

A.8   Walk-Through for EMWAVE Program

The program begins by showing the animation of a plane polarized wave.

- **Use the arrow keys to view the wavetrain from different points of view.** These keys change the altitude and azimuth angles of the viewpoint relative to the reference frame plotted with the wave. The wave is assumed to be travelling in the $+x$ direction.

- **Use hot keys F2 and F3 to change the animation speed.** These keys vary the time interval between animation frames.

- **Change the polarization state of the wave.** The quick and easy way to do this is to choose the circular polarization option under the CHANGE menu. Alternatively, you may want to reset the Stokes parameters individually using the **Stokes' Parameters** option under **CHANGE** to produce a variety of wave polarizations.

- **Observe the effect on the wave of reflection and transmission at a material interface.** Do this by choosing the Wave on Interface option under the **SHOW WHAT** menu. Press **Enter** to accept the default values for permittivity, permeability and conductivity to simulate the case of partial reflection and partial transmission into a lossy material. Note the partially standing wave that results from the superposition of incident and reflected waves. Try large values of the conductivity and a circularly polarized incident wave to examine a standing, circularly polarized wave.

- **Choose the Wave + Waveplate option under SHOW WHAT menu.** The default configuration shows the conversion of the wave from plane-polarized to circular-polarized. The hot key labeled **Rotate** allows you to see what happens for other orientations of the waveplate axis relative to the direction of plane polarization. A quick **CHANGE** to circular polarization of the incident radiation can be made from the menu to show the reverse process.

- **Try other options under the SHOW WHAT menu.**

Index

Limited Use License Agreement

This is the John Wiley & Sons, Inc. (Wiley) limited use License Agreement, which governs your use of any Wiley proprietary software products (Licensed Program) and User Manual(s) delivered with it.

Your use of the Licensed Program indicates your acceptance of the terms and conditions of this Agreement. If you do not accept or agree with them, you must return the Licensed Program unused within 30 days of receipt or, if purchased, within 30 days, as evidenced by a copy of your receipt, in which case, the purchase price will be fully refunded.

License: Wiley hereby grants you, and you accept, a non-exclusive and non-transferrable license, to use the Licensed Program and User Manual(s) on the following terms and conditions:

a. The Licensed Program and User Manual(s) are for your personal use only.
b. You may use the Licensed Program on a single computer, or on its temporary replacement, or on a subsequent computer only.
c. You may modify the Licensed Program for your use only, but any such modifications void all warranties expressed or implied. In all respects, the modified programs will continue to be subject to the terms and conditions of this Agreement.
d. A backup copy or copies may be made only as provided by the User Manual(s), but all such backup copies are subject to the terms and conditions of this Agreement.
e. You may not use the Licensed Program on more than one computer system, make or distribute unauthorized copies of the Licensed Program or User Manual(s), create by decompilation or otherwise the source code of the Licensed Program or use, copy, modify, or transfer the Licensed Program, in whole or in part, or User Manual(s), except as expressly permitted by this Agreement.
   If you transfer possession of any copy or modification of the Licensed Program to any third party, your license is automatically terminated. Such termination shall be in addition to and not in lieu of any equitable, civil, or other remedies available to Wiley.

Term: This License Agreement is effective until terminated. You may terminate it at any time by destroying the Licensed Program and User Manual together with all copies made (with or without authorization).
   This Agreement will also terminate upon the conditions discussed elsewhere in this Agreement, or if you fail to comply with any term or condition of this Agreement. Upon such termination, you agree to destroy the Licensed Program, User Manual(s), and any copies made (with or without authorization) of either.

Wiley's Rights: You acknowledge that the Licensed Program and User Manual(s) are the sole and exclusive property of Wiley. By accepting this Agreement, you do not become the owner of the Licensed Program or User Manual(s), but you do have the right to use them in accordance with the provisions of this Agreement. You agree to protect the Licensed Program and User Manual(s) from unauthorized use, reproduction or distribution.

Warranty: To the original licensee only, Wiley warrants that the diskettes on which the Licensed Program is furnished are free from defects in the materials and workmanship under normal use for a period of ninety (90) days from the date of purchase or receipt as evidenced by a copy of your receipt. If during the ninety day period, a defect in any diskette occurs, you may return it. Wiley will replace the defective diskette(s) without charge to you. Your sole and exclusive remedy in the event of a defect is expressly limited to replacement of the defective diskette(s) at no additional charge. This warranty does not apply to damage or defects due to improper use or negligence.
   This limited warranty is in lieu of all other warranties, expressed or implied, including, without limitation, any warranties of merchantability or fitness for a particular purpose.
   Except as specified above, the Licensed Program and User Manual(s) are furnished by Wiley on an "as is" basis and without warranty as to the performance or results you may obtain by using the Licensed Program and User Manual(s). The entire risk as to the results or performance, and the cost of all necessary servicing, repair, or correction of the Licensed Program and User Manual(s) is assumed by you.
   In no event will Wiley be liable to you for any damages, including lost profits, lost savings, or other incidental or consequential damages arising out of the use or inability to use the Licensed Program or User Manual(s), even if Wiley or an authorized Wiley dealer has been advised of the possibility of such damages.

General: This Limited Warranty gives you specific legal rights. You may have others by operation of law which varies from state to state. If any of the provisions of this Agreement are invalid under any applicable statute or rule of law, they are to that extent deemed omitted.
   This Agreement represents the entire agreement between us and supercedes any proposals or prior Agreements, oral or written, and any other communication between us relating to the subject matter of this Agreement.
   This Agreement will be governed and construed as if wholly entered into and performed within the State of New York.
   You acknowledge that you have read this Agreement, and agree to be bound by its terms and conditions.